The Art of
Japanese Sword Polishing

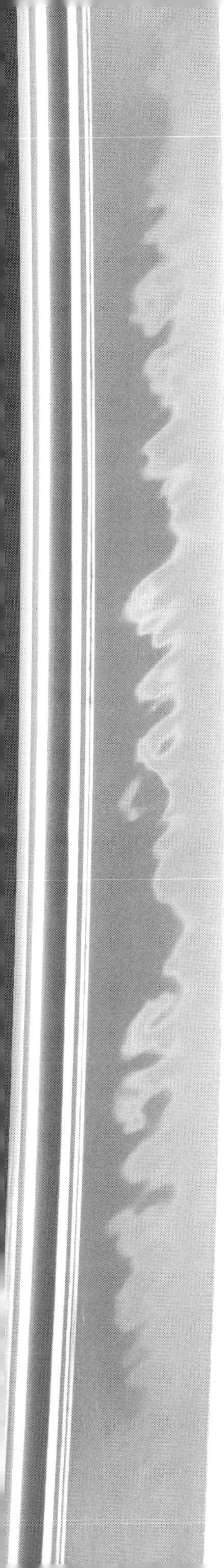

The Art of Japanese Sword Polishing

Setsuo Takaiwa
Yoshindo Yoshihara
Leon and Hiroko Kapp

KODANSHA INTERNATIONAL
Tokyo • New York • London

TERMINOLOGY

An extensive list of sword terms appearing in this book, as well as other useful vocabulary in general use by in those the field, can be found in the illustrated Appendices, beginning on page 167.

The publisher wishes to express its gratitude to all of the following people and institutions for their assistance in acquiring and/or graciously granting permission to use the photographic images in the present volume: Kawagoe City Museum (page 42), Kita-in (page 42), Saitama Prefectural Museum (page 41).

NOTE: Japanese names appear in the text following the traditional order, surname followed by given name. However, on the jacket and the first seven pages of this book names are listed in the Western manner.

Jacket photograph and photographs in Profiles by Tom Kishida
Photographs in parts I and II by Yoshikazu Yoshihara

Distributed in the United States by Kodansha America, Inc., and in the United Kingdom and continental Europe by Kodansha Europe Ltd.

Published by Kodansha International Ltd., 17–14 Otowa 1-chome, Bunkyo-ku, Tokyo 112-8652, and Kodansha America, Inc.

ISBN-13: 978–4–7700–2494-7
ISBN-10: 4–7700–2494-0

First edition, 2006
15 14 13 12 11 10 09 08 07 06   10 9 8 7 6 5 4 3 2 1

Library of Congress Catalogue-in-Publication Data available

www.kodansha-intl.com

Contents

PREFACE

The Japanese sword has long been valued for practical and aesthetic reasons. The sword as we know it today was developed during the thirteenth and fourteenth centuries. Although its original use was as a weapon, it has been appreciated as an art form since the earliest times.

A swordsmith can spend a large amount of time making a sword, but finishing the sword, refining and bringing out its final shape, and polishing it so that all the details of the steel and its pattern are clearly and easily visible is the responsibility of another craftsman: the sword polisher. There is evidence that already a lot of effort was being put into polishing swords in the Kamakura period (1185–1333).

Today, the first one-third of the polisher's efforts will leave the blade in a completely functional condition, which means that most of his time and effort is devoted to bringing out the aesthetic properties of the sword so that it might be examined and appreciated by anyone who is interested.

Japanese sword polishing has evolved over the centuries, and the earliest efforts probably concentrated on producing a functional, combat-ready weapon. However, aesthetic values have always been important to the Japanese, and historical evidence suggests that, even eight hundred years ago, efforts were made to polish swords in such a way as to make their aesthetic appeal apparent.

Polishing methods have steadily improved since the Kamakura period, many of the improvements having taken place since the start of the Meiji era (1868–1912). Beginning in 1876, when it became illegal to wear a sword in public, the aesthetic properties of the sword grew steadily in importance. The sword polishers responded to this development by greatly improving polishing methods, and today sword polishers can bring out and enhance the best properties of almost any sword.

The aim of this book is to explain how the Japanese sword is polished today, and to illustrate the methods, materials, and tools used for this process. By so doing, this volume reveals the complexities of the polishing process, and the skills and talent needed to become a competent sword polisher. Japanese sword polishing is an old Japanese craft, and an art in itself. That the polisher's work can be set forth in detail in these comparatively few pages is deceptive, in that it takes years of training and practice for someone to master the relevant skills and so be able to produce a high-quality polish that will enhance the appearance of a sword.

Setsuo Takaiwa
Yoshindo Yoshihara
Leon and Hiroko Kapp

Part I

The Polisher as Artisan

Chapter 1 INTRODUCTION TO THE POLISHER'S CRAFT

Japanese swords have long been prized for their aesthetic beauty as well as their functional properties. As polishing plays a vital role in determining the sword's final shape and appearance, an understanding of the polisher's art is essential for anyone interested in this unique weapon.

Even after the swordsmith has finished his work, a lot remains to be done: the blade must go to a polisher, a collar maker, and a scabbard maker. But it is the role of the polisher that is critical to refining the shape of the blade and producing a finish that will bring to the surface the sword's properties.

Japanese sword polishing is almost as complex as swordmaking, and describing the process is difficult. The sophistication of Japanese swords and their steel goes far beyond what is found in modern Western blades. It takes time and study to understand how these blades are designed and forged. Much of the information can be gleaned from the details of a skillfully polished blade. If what the smith has done is to be appreciated, the finish on the blade's surface must be so fine that all of the minute variations and crystal structures are clearly visible. But such an extremely fine surface finish is surprisingly difficult to accomplish. Japanese swords had matured into the form we know today by the Kamakura period (1185–1333). For this to have happened, the smiths must have been able to examine their work in detail, which suggests that, even seven hundred years ago, the craft of sword polishing already must have attained a level of sophistication.

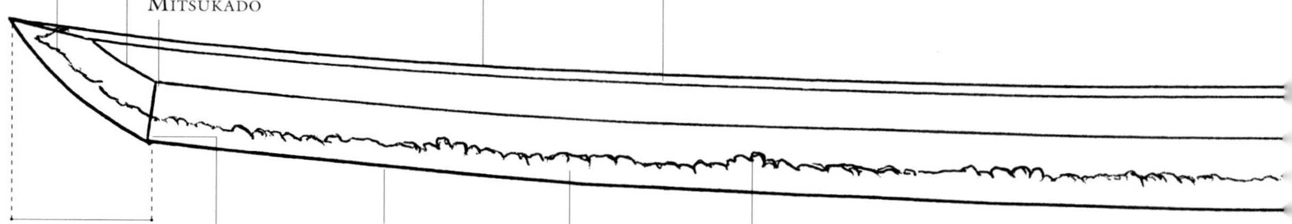

The parts of the Japanese sword. For definitions, see the Appendices, page 171.

Refining the sword's shape and producing a sharp edge is relatively simple, and it would seem that most swords intended for combat required little more than a basic polishing. However, in order to evaluate the quality of a smith's work, a much more time consuming and careful polish is required. Approximately one-half to two-thirds of a polisher's efforts are devoted to bringing out all of the details (some of which are very subtle) in a sword's steel for aesthetic appreciation and appraisal.

Since this process is time consuming and difficult, polishers often apprentice for up to ten years. The polishing process itself involves using a series of increasingly finer polishing stones—often ten or more—to highlight all of the details of the steel. This method of finishing is fraught with hazards since, if any one stage of the polishing process is performed incorrectly, the end result will be unsatisfactory. It takes years of training and experience, and the polisher must know at each stage what the steel should look like and what details should be apparent before moving on to the next stage. If a particular stone proves ineffective on a sword, the polisher must replace it. Polishing, with its numerous stages, cannot be performed by rote and be expected to result in a good finish. At each stage, the polisher must decide if the stones are producing the proper effect or finish and, if not, decide what other stones to try. Since swords are made of different types of traditional hand-forged steel comprising various ores, and an array of polishing stones exists to accommodate the variations, it requires considerable training and experience to produce the best possible finish for

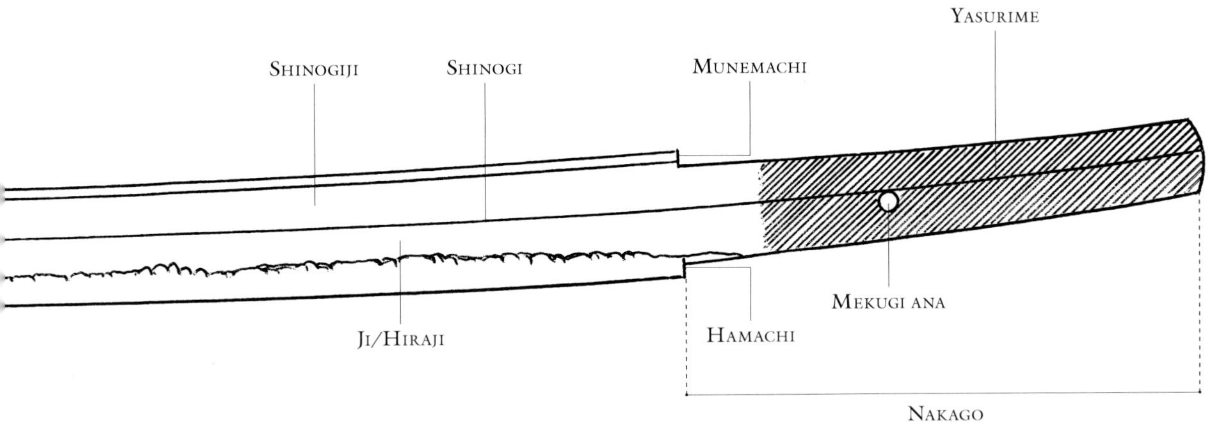

a given sword. Western blades do not receive this type of careful finish. They are typically burnished or oxidized, either of which process would virtually erase those details of Japanese blades cherished by collectors.

The Sword and Its Properties

In order to understand the task of the polisher, it is necessary to understand something about the Japanese sword and how it is made. The following provides enough background information about Japanese swords to enable both general readers and collectors to appreciate the role of polishing, as well as the problems and tasks facing the polisher.

THE SHAPE (*SUGATA*)

Each sword has its own shape, of which there are many. After the smith has forged the blade, he adds some final touches to the shape using a drawknife and a coarse file. The blade then goes through a process called *yakiire*, in which the blade is heated and then quenched in cold water to produce a hardened cutting edge with a visible pattern (*hamon*). The smith then adjusts the shape and curvature of the blade with a hammer, files, and a grinding wheel. At this stage, some smiths will give the sword to a polisher to finish, but others will proceed to complete the early polishing stages themselves, since the stones used can alter the shape of the sword.

No matter the stage at which the polisher begins his work, the shape and character of the sword must be preserved and enhanced. The main aspects of the blade's shape that should be observed are the length, the curvature, the width and proportions, the degree of tapering from the base to the point of the blade, the shape and definition of the area around the point, the cross-sectional shape, and the thickness of the blade.

All of the lines on the blade must be straight and true, crisp and sharp. The point area is often set off from the body of the blade by a line called the *yokote*, which runs perpendicular to the length of the blade (note that not all blades have a yokote line or a separate, defined point area). The curvature of the cutting edge, particularly near the point, is very important and is easily altered should a polisher grow careless. The surfaces of the blade must be uniform, smooth, and flat (or correctly curved), while the lines and surfaces must be properly formed for the blade's shape to stand out with distinction.

The Steel (*Jigane*)

The Japanese sword is made from high-carbon steel called *tamahagane*, which resembles meteorite iron and is a mixture of iron and various grades of steel. In Japan, iron ore is usually mined in a sand form called *satetsu*, and is converted into high-carbon steel in a charcoal-fueled smelter called a *tatara*.

The smith's first task is to select the tamahagane and hammer it out into a flat sheet, before breaking it into small waferlike pieces approximately 1 inch by 1 or 2 inches (2.5 cm by 2.5 or 5 cm) in size. Next, the pieces are stacked, heated, and hammered together to fuse them into a billet. This billet is then alternately hammered out and folded over a total of six times. To make a sufficient amount of steel for a sword, this piece is combined with a similar piece by means of the hammering and folding process. Pieces are combined and folded six to eight more times to produce the final steel for a sword.

Forging the steel and folding it repeatedly accomplishes a number of things. The process breaks up impurities and large crystals and works them out of the steel. At the same time, the carbon in the steel is distributed more evenly, and the resulting steel grain is much finer and more uniform. Forging takes place at temperatures that are never high enough to melt the steel, resulting in a visible pattern, or damask, on the steel surface. The pattern is formed by the oxides at the boundaries between the wafers that were stacked at the beginning of the hammering-out process.

The prominence of this pattern, or *jihada*, and its appearance depends on the smith, who controls this aspect with his forging techniques. The jihada may be prominent, resembling a network of fine lines over the blade surface, or so fine that it is barely visible. This is determined by the smith's selection of tamahagane and pattern of folding the steel and hammering, as well as by the forging temperature. The most common jihada are a straight grain (*masame-hada*), a wood-grain (*itame-hada*), and a denser, fine burl-like grain (*mokume-hada*; see page 180). The same forging process produces a characteristic color and texture (also called *jigane*) on the steel surface. The jigane varies from sword to sword and from smith to smith; it is an important characteristic of the sword and can often be used to determine the age and maker of a blade.

One additional variable affecting the final appearance of the sword is the carbon content of the steel bars. If the composite bars used to make the billet have a very similar carbon content, the jihada patterns fall into the categories mentioned in the previous paragraph.

However, if the bars differ in terms of carbon content, the layers with a higher carbon content will be harder, and the alternating layers create a contrast when polished, in the end producing a strong visual pattern resembling that of Damascus steel. These patterns will be similar to the jihada patterns described above but, whereas the traditional jihada pattern is produced by the oxide boundaries between the steel layers, the Damascus-like patterns are produced by contrasts between the steel layers themselves.

Although a jihada pattern is always visible in the work of a skillful polisher, producing a pattern by using different types of steel was only popular at certain times in Japanese sword history, and is often seen in blades produced by the Gassan School and others late in the Edo period (1600–1868).

The Hamon

One of the most important and unique components of the Japanese sword is the pattern, or hamon, formed on the hardened cutting edge.

The smith's first concern in making a sword is to produce the best steel that he can. The blade is then forged. A given piece of steel can possess different physical properties, depending on how it is treated. For the Japanese sword, having different properties within the same sword blade is essential. A good sword must have a cutting edge composed of a hard steel. However, if the entire sword were as hard as the cutting edge, the blade would be brittle and prone to cracking when used. The Japanese swordmaker overcomes this problem by combining a hard cutting edge with a softer, more flexible body. The cutting edge is composed of a form of hard carbon steel called martensite, while the body of the sword relies on much softer steel called ferrite and pearlite.

The method used to produce these different forms in the same piece of steel is ingenious, yet simple: a complex but thin layer of clay is used to coat the edge of the sword, and a thick uniform layer on the body of the blade. The blade is then heated to a high temperature and cooled rapidly by plunging it into a tank of water. The edge cools very rapidly, the steel taking the hard form of martensite. The body of the blade, with its thicker clay coating, cools more slowly, and the steel remains in the ferrite and pearlite forms. In this process, the hamon emerges, clearly defined against the body of the blade.

The hamon can display any of a large number of designs, with many complex details. The final pattern depends on many variables: the carbon content of the steel; the clay pattern used to cover the blade; the temperature to which the blade is heated; the length of time the blade is held at the high temperature before it is quenched; the evenness of heating, and so on. An experienced smith controls all of these variables to produce the desired hamon.

Functionally, all hamon should be wide enough to allow for repolishing and repairs after damage and wear. In addition, they should not cover an uninterrupted expanse of hard steel, since a strong impact could chip off large pieces. To prevent this, the clay coating placed over the edge area before heating is not simple or uniform. Thick, narrow strips of clay extend into the edge region, with its thin layer of clay. This clay pattern results in thin strips, called *ashi*, of unhardened steel penetrating the hard steel where the hamon is and extending to the edge of the blade. This can produce a complex pattern and helps lock the hardened steel of the hamon's edge into the softer body of the blade. Any chip that forms on an edge made this way will be no wider than the space between two ashi. Ashi may not always be visible, or may only be visible near the top of the hamon, but they are present in any properly made Japanese sword.

The actual hamon pattern is usually defined by a clearly visible white line. The transitional area between the hamon and the body of the sword is called the *habuchi* and is usually visible as a clear white line that defines the shape of the hamon. This habuchi is usually formed of fine crystals called *nioi* which appear where the martensite mixes with the ferrite and pearlite. The nioi cannot be seen individually, while the larger crystals that can be clealy distinguished as discrete bright spots are called *nie*. The habuchi may be made of nioi, nie, or a mixture of the two.

During the forming of the hamon a number of other effects may appear in the body of the blade or in the area (*ji*) between the cutting edge and the ridge line. For example, if the blade is heated to very high temperatures before quenching, dots or islands of martensite will form in the ji. These are called *ji-nie*. Another effect called *kinsuji* occurs in the area of the habuchi when some layers of steel have a higher carbon content than the intervening or adjacent layers. These areas of higher carbon form bright streaks of martensite and will show up clearly against the darker steel in the body of the ji. A swordsmith can produce this effect by combining steel blocks with different carbon contents during forging.

Other effects, such as *utsuri* and *chikei*, also occur in the ji. These nonmartensite effects are a result of the steel taking on different structural or crystal forms during the heating and rapid cooling. A good smith can produce these effects, and in a well-polished sword in good condition, all of these effects will be clearly visible.

The Role of Polishing and the Polisher

By now it should be quite clear that the polisher plays a pivotal role in finishing the sword. His first task is to refine the shape of a new sword (or restore the shape of a neglected or damaged sword). In either case, the polisher begins with coarse stones, aware that their careless use can easily lead to permanent damage. With these rough stones, he can remove a large amount of steel to complete the shape of the sword (or restore the blade of an older-sword to its original shape). All the blade's lines must be true, straight, crisp, and sharp, including the *shinogi*, or ridge line, which typically runs the length of the blade; the *mune* or *iori*, the line running along the back of the blade; and the yokote, which defines the *kissaki* or point area. In addition, the edge must be straight and true, with a proper taper running from the tang to the point. The point area must be shaped properly; this area is difficult to work on because it curves from the cutting edge up to the back edge of the sword and from the yokote toward the tip of the point. Further, the edge in the point area must have a uniform, continuous curve from the body of the sword to the tip of the point. Finally, the sword has a particular cross-sectional shape, from the edge to the back surface, given to it by the smith, and this should be preserved and restored: the sides of the blade may describe a straight line from the shinogi down to the edge, or the sides may be convex between the shinogi and edge.

All the lines and surfaces must be enhanced and preserved (or repaired) by the polisher. He must visualize the intended shape planned by the smith in a new sword (or the original shape in an old sword). A careless polisher can easily flatten the convex sides of the ji, dull the ridge line, ruin the graceful curve of the point, and leave a wavy rather than straight and true surface on the ji. Any errors could permanently alter the shape of the sword or require that a large amount of metal be ground off from other parts of the sword to restore balance.

The polisher's other goals are to finish the surface of the sword so that all the details are visible. This requires a pristine finish completely free of scratches. If the quality of the finish is fine enough, the jihada and the steel's color and texture will be clearly exhibited. With a good polish, a great deal can be determined. Is the jihada composed of fine or coarse lines? Did the smith mix steels of different carbon content to produce an additional surface pattern? Does the jihada have an itame-hada, a masame-hada, or some other pattern? Do the steel's color and jigane indicate it comes from a particular period or place? Are special features of the sword visible? If present, they should be clearly defined.

If the polisher has done his work well, the hamon and its characteristics must also be apparent. One should easily be able to see if the hamon is formed of nie or nioi. The nioi line should be continuous and unbroken. Details in the hamon such as ashi or kinsuji should also be clearly visible, and the pattern of the hamon should be easy to discern and identify.

The hamon does not always appear brighter than the ji. The polisher must use his skill and techniques to bring out the nioi and nie to produce a clear, well-defined hamon that contrasts well against the ji. This is not always easy, and a good polisher must use his judgment in deciding how best to achieve a worthy finish.

In summary, good polishers coax out and enhance all of the details of the blade—its shape, the jigane, and the hamon—so that the sword's features are apparent. All of these details are necessary for a complete and accurate appraisal of the blade. With close observation, one can often determine the period in which a blade was made and its style; the city or town of origin; and sometimes, even the individual smith. Because the craft of polishing requires a good eye for detail, good judgment regarding the appropriate methods and techniques, and a wise selection of polishing stones to be used, it should come as no surprise that the most prominent appraisers in sword history have usually been sword polishers. Such is the importance their role.

Chapter 2 THE POLISHER'S MATERIALS AND METHODS

Before diving into a detailed explanation of the methods and techniques used by Japanese sword polishers, it is necessary to outline the basic polishing process and the materials used. What follows is an overview of how polishing is carried out, designed to familiarize the reader with the stages and terminology. Traditionally methodical and detailed, the process can engage a careful polisher for up to two weeks when the Japanese sword is new.

The overall process is divided into two main stages called foundation polishing (*shitaji togi*) and finish polishing (*shiage togi*). New swords that are polished for the first time are subjected to all of the steps described here. However, older swords can be repolished beginning with whichever step is appropriate. For example, if the sword is deeply rusted or chipped, the polisher might start with the first series of stones, the *arato* or *kongoto* or *binsui*. If the sword is in fairly good condition, with a cloudy and indistinct surface but little or no rust, the polisher might start with a *chu-nagura* stone. A good polisher will start with the finest grade of stone possible to remove the old finish. Starting with a needlessly coarse stone will result in removing more metal than is necessary. Such consideration for a sword is important; the large number of five- and six-hundred-year-old swords still extant indicates how long these swords can last if they are properly maintained and repolished. Careless or thoughtless polishing and excessive removal of metal during repolishing can greatly shorten the life of a blade.

Among the natural polishing stones used for the foundation polishing are (from left) uchigumori ha-to, uchigumori ji-to, koma-nagura, chu-nagura, kaisei, and binsui. The uchigumori stones are the finest, the binsui stone the coarsest of these.

Synthetic stones used for foundation polishing. Stones, from left: nagura, kaisei, kasisei, binsui, and arato.

Foundation Polishing

In foundation polishing, large stones are used and the sword is moved over the stones. Seven stones or more may be used during this process, depending on the sword, its condition, and how well the blade responds to the stones available to the polisher.

Specific goals should be accomplished at each stage before the polisher progresses to the next stone. After polishing with a particular stone, if the finish is not good enough or if proper quality for that stone is not achieved, the final polish will not be satisfactory.

The stones are listed here in their order of use. They form a continuous spectrum of increasing hardness and of increasingly fine grain structure. Traditionally, all were either natural sandstones or limestones; however, today the coarser grades are usually synthetic, while those used later are natural stones. In general, a polisher might use the synthetic stones up to the *kaisei* stage, and natural stones from the *nagura* stage onward.

Arato (literally, "rough stone") is the first stone to be used. It is very coarse and is used only for finishing new swords or for repairing very badly rusted or damaged swords. When using this stone, the polisher moves the blade over the stone in a direction perpendicular to the blade's length.

Kongoto is another very coarse stone that is somewhat finer than arato. It can be used for badly rusted or damaged swords, or to remove the scratch-like markings made by the arato. As with the arato, the polisher moves the blade over the stone in a direction perpendicular to the blade's length.

Binsui is finer than kongoto. It can be used as a first stone for polishing if the condition of the sword is not extremely bad, but otherwise it is used to further polish the blade, as well as to remove the markings left by the kongoto stone. The part of the blade with the kongoto marks is held at an angle of fifteen degrees to the binsui. The polisher must use the binsui until all the earlier marks have vanished. This finish must be uniform over the entire blade. The difference in the angle helps to ensure that all of the remaining marks are from the binsui and not the kongoto. If the marks left from the previous stone are not removed now with the binsui stone, they will be much more difficult to remove with the succeeding stones. This is true for each stage: the polisher must always be careful to remove all marks from the previous stone.

The next stone, kaisei, is still finer, but coarse enough to damage a blade if used improperly. At this stage, the polisher moves the blade over the stone diagonally. For most swords, this is when the hamon pattern starts to emerge.

Chu-nagura follows kaisei. In general, from this stone on, all are natural stones. With the chu-nagura, the blade is moved lengthwise over the stone. Again, the blade must be left with a uniform finish and all diagonal marks from the kaisei stone must be removed. The hamon should become more visible but the ji will exhibit only a whitish finish. The marks left by the chu-nagura are very shallow and appear more like rough smudges than scratches. The area from the yokote to the tip of the point is polished in a direction perpendicular to the length of the blade. This area is polished in the same direction with all of the stones, regardless of the treatment given the rest of the sword.

The next stone is *koma-nagura*. It is similar to the chu-nagura and not much finer. The blade is again moved lengthwise over the stone. The marks from this stone appear as smudges. There is usually not much difference in the finish between that of the chu-nagura and of the koma-nagura, only that the finish will be a little finer, the surface clearer, and the hamon clearly visible.

Suita is used next. The sword is again moved lengthwise along the stone. This is a hard, light blue-gray stone and the hamon must be completely clear and scratch-free after it has been used. The transition from the nagura stones to the suita is difficult and can take considerable time. The steel's surface should become a bit darker and there will be no visible scratches; the hamon should be easily visible. The suita is a type of *uchigumori ji-to* stone. It is only used to polish the *ha* before the uchigumori ha-to is used. Since it is coarser than the rest of the uchigumori stones, the polishing time is shorter.

After the suita, the next two stones used are *uchigumori*: uchigumori ha-to and *uchigumori ji-to*. These stones are also a blue-gray color, but generally the ji-to will be darker than the ha-to stone. Again, the sword is moved lengthwise over the stones. The marks left by these stones appear as very fine lines or smudges, not like scratches at all. The smudges should be parallel. Some describe the marks as lines made by fine noodles. As each uchigumori stone is used, the contrast between the hamon and the ji should increase, and the details of the hamon—its nioi, nie, ashi, and other details—should stand out, clear and bright in the habuchi. After using the ha-to stone, the hamon should be free of marks or smudges. After using the ji-to stone, the visible surface pattern (jihada) and details of the ji should be clear. There should be no scratches or any visible marks of any kind left anywhere on the sword's surface, since the work in the following stages will use such fine stones

► Polishing stones displayed at Nomura's, a shop of long standing in Asakusa, one of the oldest districts in Tokyo. Most of the stones on the shelves are natural stones and the ones in the foreground are synthetic.

SHARPENING STONE

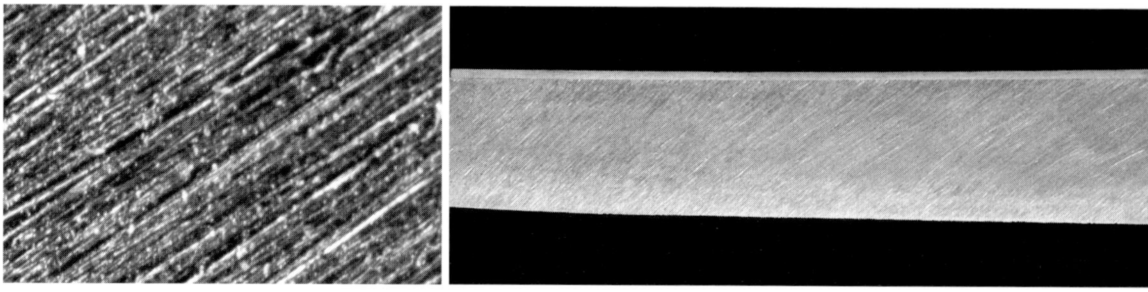

Markings from a synthetic arato stone.

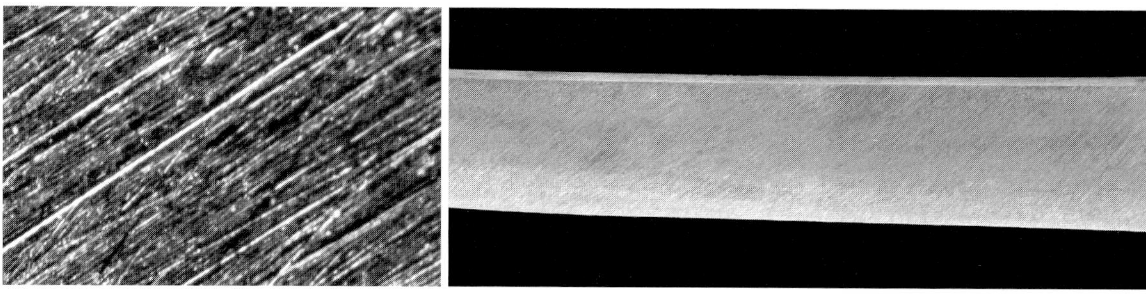

Markings from a synthetic omurado stone.

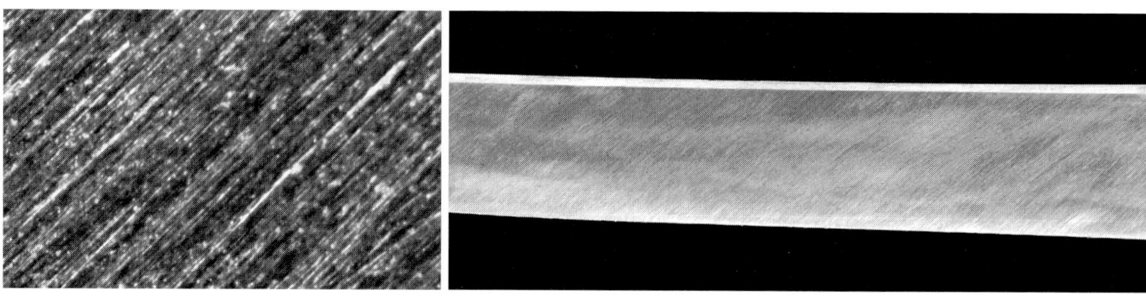

Markings from a synthetic binsui stone.

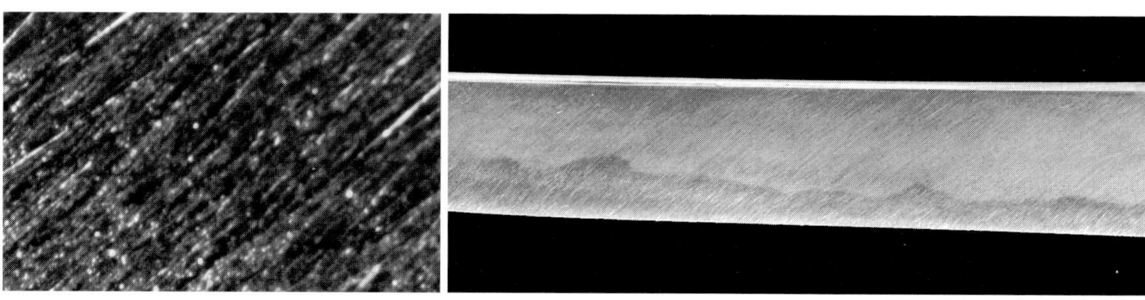

Markings from a natural binsui stone.

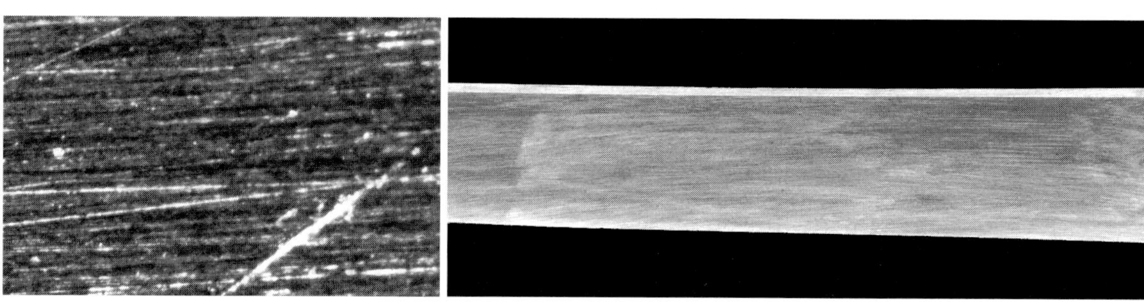

Markings from a synthetic kaisei stone.

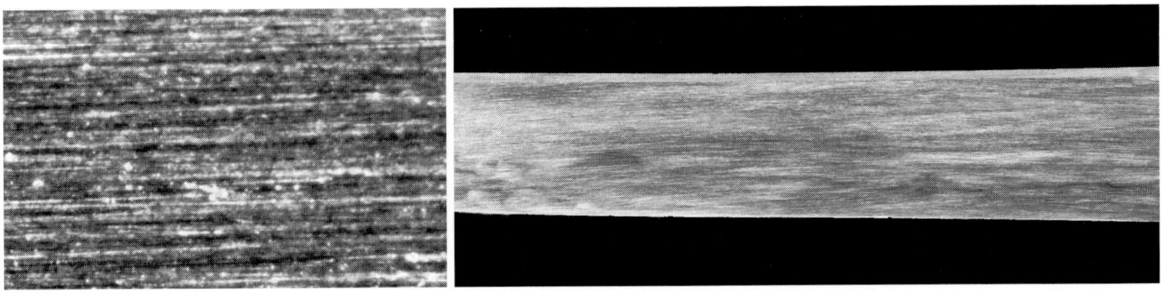

Markings from a natural kaisei stone.

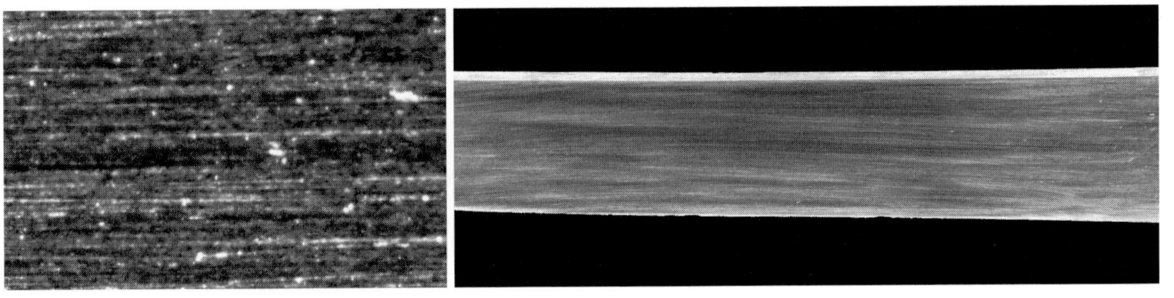

Markings from a natural nagura stone.

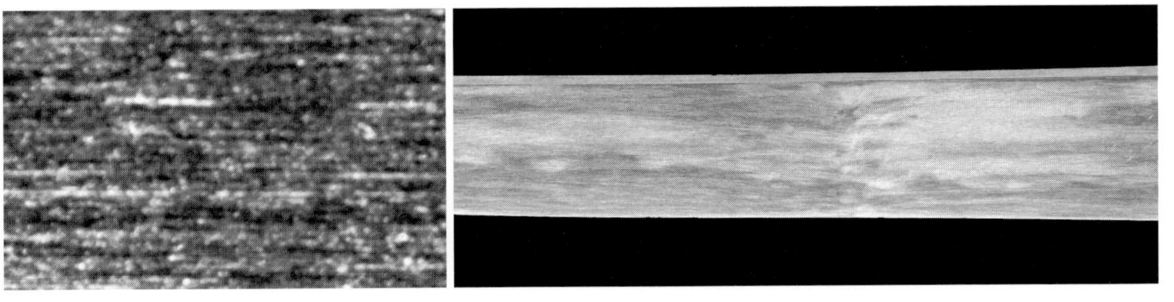

Markings from a natural koma-nagura stone.

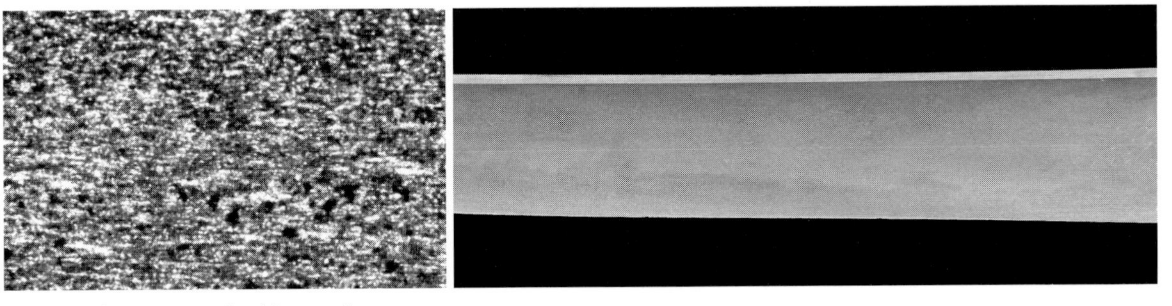

Markings from a natural uchigumori stone.

and materials that it will not be possible to remove any remaining scratches or marks.

Finish Polishing

Finish polishing begins after all of the large stones have been employed. In the remaining stages, the stones and tools are moved over a stationary sword, in contrast to those of foundation polishing.

The first stage of finish polishing involves the use of the *hazuya* stone to go over the entire blade. This stone is made from an uchigumori stone that has been split into thin slices or flakes, attached to a backing of Yoshino paper (see pages 101, 116, 118) with Japanese lacquer (*urushi*), and ground down on coarser natural stones until it is almost translucent. The finished stones are approximately 1/200 of an inch (0.13 mm) thick. These thin stones (or *tsuya* stones) are then cut into small squares or rectangles, approximately 3/4 to 1 inch (1.9 to 2.54 cm) in size.

Even though the last uchigumori stones that were used were very fine, the weight of the sword and the pressure of the polisher's hands will have resulted in some marks or smudges having been left. The purpose of the hazuya stone is to remove these last, very fine marks. The resulting finish will leave a whitish cast on the surface of the blade, with no visible marks of any kind remaining. If the polisher has done a good job up to this point, the hazuya stage could be completed in a short time; in fact, some polishers even omit this step.

To use the hazuya stone, the polisher holds the blade on his lap or on a well-padded clean support. He forms a fist with his four fingers, the backs of his fingers resting against the smooth back surface or ridge of the blade and his thumb projecting over the blade surface. He then goes over the blade, section by section, working from the cutting edge up to the shinogi or ridge line, and then back down to the cutting edge. Only about one- to two-inch sections are polished each time, and the polisher works his way down the blade from the tang to the yokote, but does not polish the point area.

After the hazuya stone, a series of *jizuya* stones are used. These are prepared exactly like the hazuya but are made from a finer type of stone called *narutakido*. They range in color from a yellowish brown to a reddish brown. These stones are also ground down until they are very thin and translucent. With the jizuya stones, the polisher once again goes over the entire blade, section by section, using the stones in the same manner as the hazuya. Usually only the ji is polished with the jizuya stones. The kissaki, mune, and shinogiji are not polished with the jizuya.

There are two styles of jizuya stones used depending on the polishing tradition. The Fujishiro school uses jizuya stones backed with Japanese paper and lacquer. A grid of cracks is made in the stone to allow flexibility, permitting the jizuya stone to closely follow the sword's surface and produce a uniform final polish. The Hon'ami school makes the stones very thin and translucent, but does not back them with paper and lacquer. Instead, after they are thin enough to use, the stones are broken by forcing them down on a gently curved surface. This creates a group of very small stones that the polisher keeps together under his thumb while working them over the sword exactly as he would the hazuya stones. This permits the small stones to closely follow the sword's surface, including any irregular areas, and produce a uniform final polish.

Takaiwa is polishing the ji with a piece of jizuya stone perpendicular to the lengthwise axis of the sword.

In general, whichever style is used, the polisher will employ three jizuya stones: after working over the surface of the entire blade with one stone, the process will be repeated twice more. Each jizuya stone varies from that used before, and generally each is thinner and harder than the preceding one.

If these last stones are properly chosen and used, the blade's surface will darken and the jihada, as well as all of the features in the ji (such as kinsuji, ji-nie, chikei, and utsuri) will become apparent. In addition, the color, texture, and jihada should have a uniform appearance over the entire surface of the blade.

The next step is the application of *nugui*. The most commonly used nugui contains very finely ground iron oxide (*kanahada*) in an oil suspension. Even after finishing the blade with the jizuya stones to bring out all of the fine details in the sword's surface, a harsh sheen remains. The nugui preparation darkens the steel and softens the appearance of the finish. It also enhances the color and appearance of the jihada.

Holding the blade in his lap or on a well padded rest, the polisher applies the nugui, filtered through paper, and dabs the blade in places several inches apart. The polisher uses a small piece of folded cotton cloth under his thumb (in the same manner as the tsuya stones) to rub the nugui over the blade, section by section. The cotton is folded into a small rectangle approximately ¾ by 1 inch (1.9 by 2.54 cm). The polisher applies the nugui at the base of the blade just above the tang, beginning at the cutting edge and works his way up to the shinogi and then back again. A 1- to 2-inch section is worked on at a time. The polisher then proceeds from the area above the tang to the point and back down again.

Several types of nugui can be used depending on the finish desired. The material used in the nugui can be iron oxide, or one of several types of stones or minerals.

The next step is to produce a brighter contrasting hamon. Although the hamon is often the same color or darker than the ji, it stands out and is clearly visible. This effect is usually achieved with a *hadori* stone. While the hadori is the same as the hazuya and is used in a similar manner, it is used only for the hamon. The blade will have acquired a relatively dark finish after the nugui stage. Using the slightly coarser hadori stone on the hamon will lighten it and produce a strong contrast with the surrounding areas. How closely the hadori finish follows the hamon will depend on the skill of the polisher as well as the nature of the hamon. In general, the polisher will try to follow the hamon as closely as possible, but if the hamon waves are very fine and irregular, the polisher will not be able to polish along its exact path. Instead, he will produce a white "apparent" hamon which will trace the outline of the real hamon as closely as possible.

Another type of finish for the hamon is the *sashikomi nugui.* In this case, the entire hamon is lightened with a hadori stone, and the hadori effect will extend into the ji. After the hadori stone is used to lighten the hamon area and the region around the hamon, a type of nugui is used. However this sashikomi nugui is prepared differently, and contains ground magnetite (lodestone) or other minerals. When used properly, sashikomi nugui will darken the ji, but not have much effect on the color or brightness of the hamon. The technique produces a white, light-colored hamon, but a dark ji, leaving the hamon clearly contrasted and visible against the body of the sword. With this finish, the appearance of the hamon follows its contours exactly.

Which hamon finish is used will depend on the properties and condition of the sword and the hamon, as well as the tastes of the owner and the polisher. The hadori finish can always be used to good effect,

Takaiwa is polishing the point area on a hazuya, which is placed on top of the narumedai.

but the sashikomi finish can generally be used only if the sword has a strong, clear nioi line with a well-defined hamon.

The point area is polished next, with the area near the *yokote* and the yokote itself polished before the rest of the point area. To define the yokote line, a bamboo mask is placed on the blade where the yokote line will be. A square piece of hazuya is then moved along the bamboo mask on the point side of the yokote using a bamboo spatula to control the movement and to put enough pressure on the hazuya for it to polish the area in the point. Proper movement of the hazuya along the bamboo mask will produce a matte white finish on the polished point, which will clearly set off the yokote line against the darker color of the ji. After defining the yokote, the rest of the point area is polished. Because of its complex shape, the point is not polished like the rest of the sword: a device called a *narumedai* is used. This is a small narrow piece of wood about two inches wide by six inches long, on the underside of which a series of lengthwise saw cuts are made. This results in a

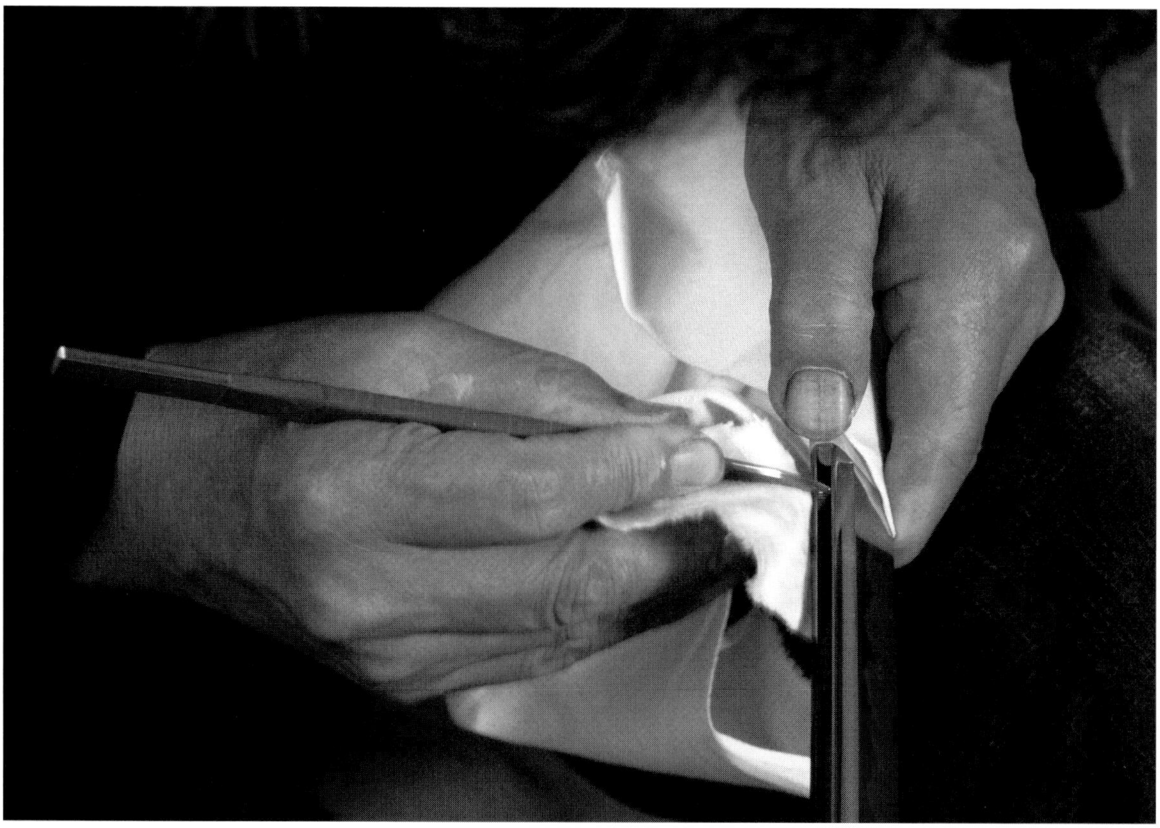

Takaiwa is burnishing the mune with a burnishing spatula, using a piece of cotton cloth placed between his fingers and the steel surface for protection and ease of movement.

flexible working surface that gives way under pressure. A piece of paper is folded to produce ten layers and placed over the narumedai; a piece of hazuya is placed over the paper, making the construct act like a springy universal joint under the hazuya, allowing it to follow all of the contours and the shape of the point. The point area of the sword is then polished between the yokote line and the tip by moving the blade in a direction perpendicular to its lengthwise axis over the hazuya. This produces an even, matte white finish.

The last step in polishing the sword is to burnish certain parts of the blade, for which the polisher uses a hardened steel needle. By rubbing it over the blade, he can produce a bright, mirrorlike finish. Traditionally, only certain parts of the blade are burnished this way; which surfaces depends on the style of the blade. In general, burnishing is used on the upper part of the blade, and between the ridge line and the back. If the blade has no ridge line, only the back unsharpened surface is burnished.

The polisher first cleans and prepares the surfaces of the blade, and then burnishes the appropriate surfaces of the blade, section by section, in a manner similar to that used with the hazuya and jizuya stones. Often, the polisher will use a second burnishing needle to go over the surface again to achieve the desired bright, clear finish. The burnishing techniques used can vary and will depend on the polisher and the sword.

Sword polishing is a unique Japanese craft. As should be clear from the above description, the aim of the polisher goes far beyond merely sharpening or cleaning the sword. The polisher's goal is to enhance the appearance of the blade and bring out the details and its individuality, so that the owner of the sword can easily see them and the sword's character.

The work involved is extremely detailed and time consuming. There is a reason for every step used in the process, and no shortcuts are possible if the best results are to be obtained. Extreme care is necessary: one error at any stage of the process can lead to a deep scratch or mark on the blade, which will require considerable time and effort to remove. The best polishers work at a surprisingly sophisticated level, constantly striving to coax forth the individual properties of each sword.

Not only are all good swords distinctive, but individual polishers—because of all the time, effort, and skill necessary to produce a good polish—produce a unique finish. So distinctive is the polisher's touch that a practiced eye can often identify the polisher from a sword's polish.

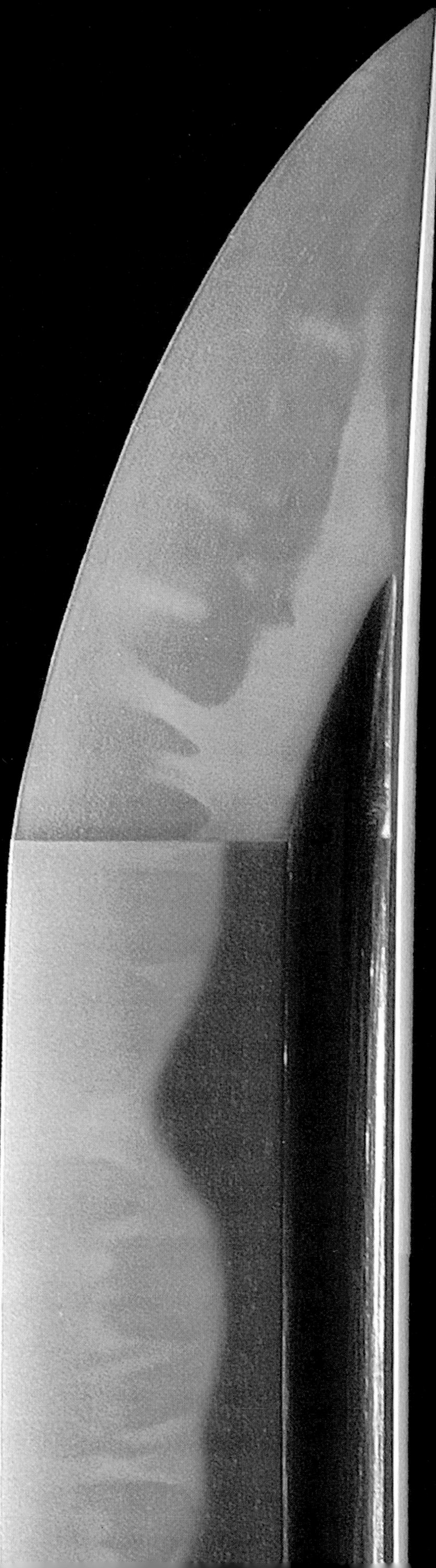

This sword by Yoshihara Yoshindo has been fully polished. The stylized hamon indicates the polisher chose the kanahada nugui method, followed by hadori.

Chapter 3 EVALUATING THE POLISH

The traditional Japanese polishing process is long and detailed. If a polisher has the requisite skills and has conscientiously followed all of the traditional polishing steps, a good polish should result. If the polisher's work seems less than satisfactory, it could well be because the sword was not properly made or it is not in good condition—having been worn down by excessive use or too much polishing that has, over time, removed too much of the sword's original steel. To effectively evaluate a sword, it often helps to examine the quality of the polish as well, since it is through the lens of a polisher's work that the quality of the sword is revealed.

In evaluating a Japanese sword, sections of the sword are examined in sequence. In general, each of the following elements are assessed discretely: the shape (sugata), the steel surface color and texture (jigane), the hardened edge (hamon), and the point (kissaki).

1. The shape. A sword should have a good clean shape with good proportions, and all of its lines and surfaces should be crisp and clear. The surfaces should be smooth and flat, or continuously curved or contoured with no waves, dips, or discontinuities.

2. The steel surface. The surface should have a clear color, texture, and appearance. It should not be shiny or featureless. Different forging methods developed in Japan over the centuries have resulted in a range of steel compositions and appearances. The distinctive features of the steel are key elements in appreciating and judging the sword. The visible surface pattern varies with individual smiths and schools and is vital in judging or appraising a sword. The surface pattern can be very prominent and clear, or subtle and hard to see.

3. The hardened edge. The edge pattern, or hamon, is unique to the Japanese sword. A good hamon should be continuous and unbroken for the length of the sword, with no gaps or breaks; it should form a recognizable traditional pattern along the edge and be clear, bright, and well defined against the body of the sword.

4. The point. The area comprising the point of the sword should have a clear, well-defined *boshi* (the hamon on the point); it should be of a size and proportion that suits the blade. The yokote line should be well-defined, straight, and clear and the entire point should be well-defined against the body of the sword.

The polish should be evaluated to determine the true value of the sword. The same four areas should be examined, as well as the final burnish.

1. The shape of the sword. A polisher should not drastically change the shape of the blade. Good work can enhance or refine the shape and appearance of a sword.

The following points should be considered when examining the polish in relation to the shape. All the surfaces of the sword should be true and flat. There should be no dips or waves in any surface on the ji, *shinogiji*, or back (mune). The shinogi and iori/mune should be uniformly crisp, sharp, and straight. Errors here might arise from improper work with the early-stage polishing stones.

2. The steel surface and appearance (jigane). Look at the ji under a strong light. There should be no traces of scratches perpendicular or diagonal to the lengthwise axis of the sword. Such scratches might be present if the polisher had not spent sufficient time with the nagura or earlier stones, or had not been careful enough with them.

The ji should be clear and details such as the jihada, color, texture, and boundary of the hamon (*habuchi*) should be clearly visible. The prominence of the jihada can be controlled by the smith and the polisher, its final appearance dictated by the effect that they wish to achieve. However, the nature of the jihada should be clear (i.e., masame-hada, itame-hada, mokume-hada, or some other form) whether its appearance is strong or subdued.

If the ji appears cloudy or indistinct, this could mean that the foundation polishing was insufficient at one or more stages: perhaps the work with the jizuya stones was rushed. It could also indicate a low-quality, mass-produced blade from any of several historical periods. The whitish effect or *utsuri*, if present, may not show up or be barely visible. A weak habuchi can indicate insufficient work at the foundation polishing stage (though it sometimes might indicate a weak hamon).

The jihada should be visible, and the blade's surface should be even and smooth. Blades in which the jihada appears harsh, whitish, and very strong may have been subject to an acid treatment to etch out the jihada at some point during the polishing process. This type of short-cut or artifice should not be necessary if the polisher has the requisite skill and training, and spends a reasonable amount of time on the sword.

3. The hardened edge. The hamon should be clearly visible against the body of the sword, with both the pattern and details distinct. This is not always possible if the hamon is weak: it could be a diffuse hamon, which could mean it was not properly made, much of it was polished out, or it was exposed to high temperatures at some time. Some types of hamon may not have a clear, well-defined nioi line because of the style in which they were made (e.g., Satsuma *shinto* blades; see page 44). In the case of a weak hamon, the polisher must use a hadori stone to enhance it. A good polisher will closely follow the discernible signs of the hamon to produce a white, clearly visible hamon that will enhance the overall appearance of the blade. If a polisher is in a hurry or not highly skilled, he might simply make an apparent, strongly visible straight (*suguha*) hamon with the hadori stone, even though the actual hamon under the hadori polish might not be suguha. A fast application of the hadori stone along a straight line above the hamon is easier than taking the time to carefully follow the actual hamon.

The hamon should be clear, with its pattern and the details (ashi, *yo*, and kinsuji; see Hataraki section in Appendices) should be easily seen. There should be no streaks, smudges, or other markings. Any such visible marks mean that a polisher rushed through one or more stages, failing to remove the residual markings left by the previously used polishing stones.

The hadori stone is employed by most polishers to finish the hamon, and while the effect of the hadori can be minimal, it can also be overdone.

For example, the hadori stone can be used to cover a large area of the ji over the area of the habuchi and the nioi line, and rubbed in so strongly that the resulting finish can overwhelm or obscure the real shape, size, and form of the hamon.

A number of other techniques can be applied to finish a hamon, and the judgment of the polisher, as well as his skills, will determine the final appearance and form of the hamon.

4. The point. The area around the point is finished after all of the other polishing steps have been completed. A bamboo mask and a hazuya stone are used to make the yokote line. The yokote should be positioned in the proper place and perpendicular to the back of the sword (mune). The yokote should be straight, clean, and continuous over the width of the ji. The polish on the point area should have an even, matte white finish and there should be no scratches. The boshi should be clear and easily visible. If it cannot be seen, or only barely, that may be due to the quality of the polish or because there is no boshi, as a result of excessive polishing or wear, or damage to the sword.

An indistinct boshi can result from insufficient time or skill at the foundation polishing stage. An improperly placed or angled yokote can result from a polisher's inappropriate decision regarding where to place it. An uneven finish around the point can result from improper use of the hazuya on the narumedai.

5. Burnishing. The shinogiji and the mune are burnished to produce a bright mirrorlike finish. This burnishing should be uniform and bright enough to produce a clear contrast with the main surface of the sword's body. There should be no spots or unevenly burnished areas. In addition, the shinogi should be sharp and well defined, and the burnished area should not exceed the accepted boundaries.

Some polishers add decorative burnishing marks, which will appear as a series of parallel burnished lines on the shinogiji under the habaki, and on the mune between the burnished area and the point. The marks originally meant that the polishing was done by a Hon'ami school polisher. Today, however, most polishers add these burnishing marks as a matter of course.

The guidelines here are specific, but may at first seem vague. If you follow them while viewing a variety of swords, you will soon be able to make the distinctions noted.

Remember, too, that proper traditional Japanese polishing is a complex task; polishers can apprentice for as long as ten years while learning their profession, and not all polishers and their apprentices are created equal. This means that, when evaluating a sword, one should consider how much of its appearance is due to the quality of the polish. Even a newly polished blade may have indistinct lines, a cloudy ji, and an obscure hamon that could dimish the value of its workmanship.

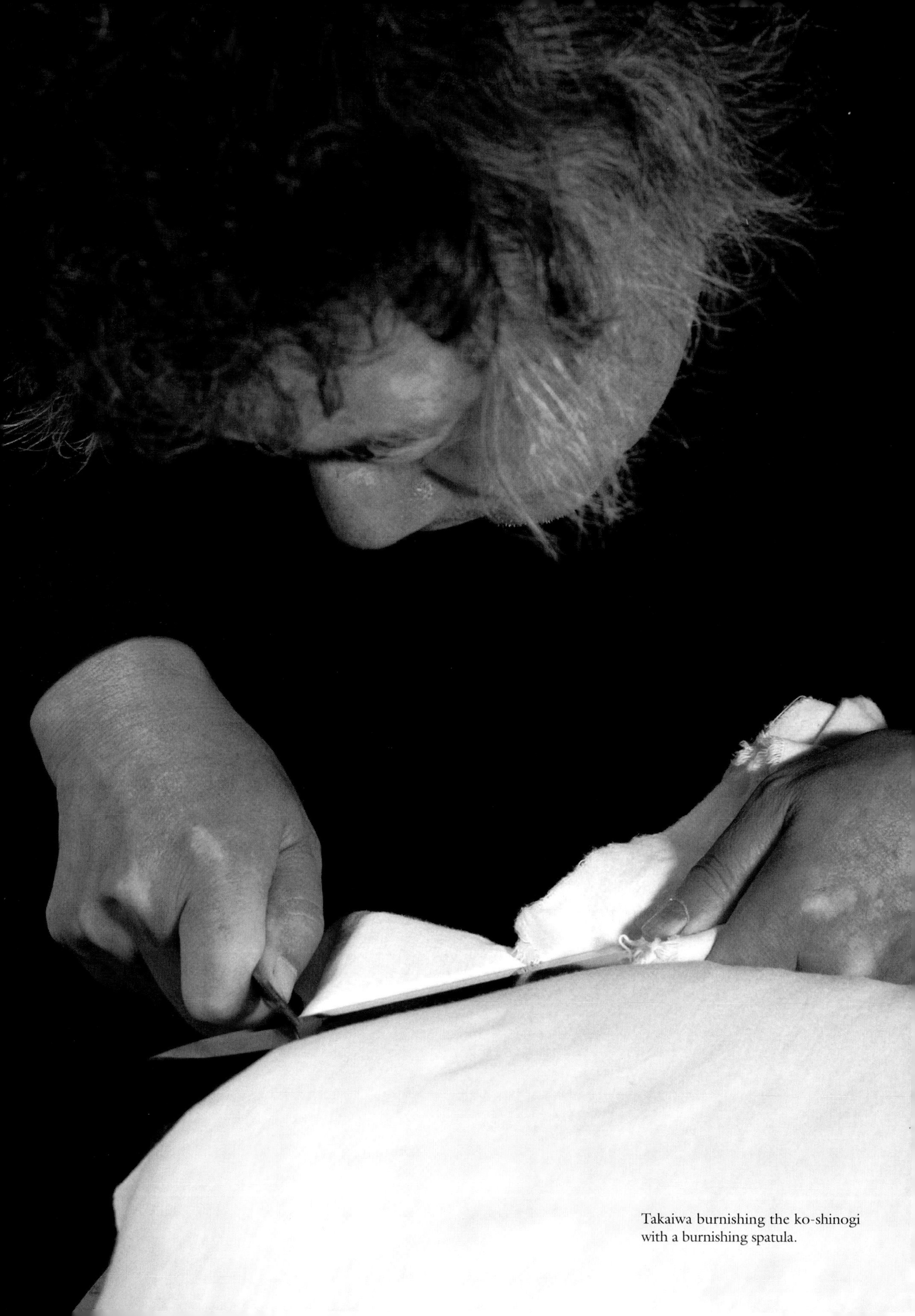

Takaiwa burnishing the ko-shinogi
with a burnishing spatula.

Chapter 4 HISTORY OF POLISHING

The sophistication of the sword polishing process suggests it has a lengthy history. However, unlike a sword, evidence of a well-conceived polishing can vanish due to rust, neglect, or just the passage of time. Since the actual work of the older polishers is no longer extant, tracing the history and development of Japanese sword polishing is a challenging task.

It seems likely that the art of sword polishing has existed at a sophisticated level in Japan for a considerable period. The Japanese have long appreciated their swords, and a large vocabulary exists to describe all of its aspects. These terms not only describe elements like the shape of the sword, which is easy to see, but the less clearly discernible details of the steel and hamon, most of which cannot be seen unless the sword has been give a quality polish. The relevant vocabulary appears to date to at least the Muromachi period (1394–1573), indicating that the Japanese were able to closely examine and appreciate their swords long ago.

The earliest stages of the polishing process refine the shape of the sword to produce a functional weapon. However, references to extremely fine details of the steel and hamon (e.g., nioi, nie, ashi, kinsuji, jihada, jigane) suggest that, at the very least, the basic methods of modern Japanese polishing techniques existed as early as the Kamakura period (1185–1333).

Further, it seems likely that the initial elements of sword polishing involved the stones that are used to shape the sword's surfaces and produce an adequate cutting edge for practical use. Dr. Honma Junji—cofounder of the Nippon Bijutsu Token Hozon Kyokai (NBTHK; The Society for the Preservation of Japanese Art Swords) and authority on Japanese swords—described evidence of sword polishing found in archeological sites that were able to shed some light on the development of sword polishing (*Token Bijutsu*, volume 39). His article discusses a sixth-century sword, excavated from a tomb, which was found with portions of its blade in good enough condition to see that it had been polished with a stone similar to the nagura, indicating that swords were already being given careful finishing work by approximately the sixth century. This particular sword is now a national treasure.

Another straight blade (*jokoto*) described in the same article was excavated in 1939 from a tomb in Shimane Prefecture. This blade was from the Nara period (710–94) and was well preserved, with sections still unrusted. Close examination indicated that it had been polished with a nagura stone. The nagura stone is used after the arato, kongoto, binsui, and kaisei stones in the foundation polishing process, indicating that blades in the Nara period were receiving careful polishes. Skilled work with the nagura stone will often bring out most of the details of the hamon. Thus, by the Nara period, polishing techniques were likely advanced enough to allow the Japanese sword to have a well-finished shape and visible hamon.

Two centuries later, the *Engi Shiki*, a collection of government documents compiled between 905 and 927, records the making of a sword. After the sword was forged, at least several polishing stones were used successively to remove forging and file marks from the steel. Next, bundles of straw were burned and the resulting alkaline straw ash was used to clean the blade, then mixed with oil to shine or burnish it. The process, which took up to three days, suggests a growing attention to blade polishing. It is hard to know exactly what the blade would have looked like, but if stones up to the nagura stage were employed, the hamon might have been quite visible (although quite narrow and straight as in existing jokoto blades). If the final stones used were fine enough, most, if not all, visible scratches would also have been removed. The cleaning or burnishing process with ash and oil probably left the blade appearing bright but without many clear surface details. A sword polished in this manner would have a well-defined shape and clean surface, but would not be suitable for detailed appreciation since few surface details, or even the hamon, would be apparent.

The jokoto swords, with their straight blades, seem to have begun evolving into the modern curved Japanese sword by the beginning of the Heian period (734–1185). In addition, the wider, more complex hamon we know today, with loops or waves (*choji*) and regular waves or semicirles (*gunome*), started to replace the narrow, straight hamon of jokoto blades. The steel surface of the sword also began to resemble that of modern blades, indicating that forging methods were evolving.

It appears that there may have been no custom of appreciating the sword during the period when jokoto swords were produced, from about the mid-seventh century to the end of the tenth century. During the tenth century, as Japanese swords were changing, it seems likely that the polishing process also evolved to accommodate the new aesthetic elements readily visible on the sword.

The illustrated handscroll known as the "Matsuzaki Tenjin'engi" (1211, Important Cultural Property) shows a military commander named Aritada admiring a sword. The scene is interesting because of its

A smith and a polisher. A detail of a painting depicting eight pairs of craftsmen in the Kamakura period (1185–1333). (Muromachi-period copy, Saitama Prefectural Museum)

detailed depiction. Aritada has drawn his sword from its scabbard and is holding the blade up to examine and admire it. The blade appears to be modern, with a curved and graceful shape. Aritada is holding the hilt in one hand and the bare steel of the upper part of the blade with the sleeve of his kimono rather than with his bare hand. This picture not only suggests that the sword was beautiful enough to admire, but also that its condition and polished finish were sufficiently good and important enough to protect from the oil and salt of the skin. His manner of handling the sword to admire and appreciate it is similar to how this would be done today.

If it was possible to appreciate a sword in such manner at that time, the technique of sword polishing must have advanced sufficiently for it to be possible to see the details of the steel and hamon. While there are no polishes left from this period to examine, the polishing technique must have advanced to a point where stones finer than nagura were employed; it seems logical that polishers must have been capable of using at least the suita and uchigumori stones by this time to produce a clearly visible hamon and jigane, otherwise it would not be possible to appreciate or examine a sword in the manner depicted in the scroll.

The oldest publication in existence about Japanese swords, the *Kanchi-in Bon* (dated February 21, 1423), appeared in the Muromachi period. The document is also referred to as the *Showa Meizukushi*, because it contains references to the year 1316 of the Showa era, which is in the Kamakura period. It was in the Kamakura period that the Japanese sword matured, assuming the shape and form we know today. One of the key events in the Kamakura period's sword history is the effort by the Emperor Gotoba (1180–1239) to perfect the Japanese sword (1207–11). The *Showa Meizukushi* lists the names of the swordsmiths who forged swords in the presence of Emperor Gotoba. Interestingly, the names of two polishers, Kunihiro and Tamesada, who worked for the emperor during the same period, are also included, an indication of the growing importance of, and need for, polishers.

The *Showa Meizukushi* also contains illustrations of sword tangs with their file marks and signatures and has comments about the hamon and other details. This means that, by the thirteenth century and early fourteenth century, details of the hamon and jigane were being observed and studied, and apparently the work of different swordsmiths compared and catalogued. This suggests that there was a high level of skill among the polishers of the time.

A Japanese dagger (*tanto*) was discovered in a wooden figure in Ankokuji temple in Hiroshima. The blade has a relatively clean steel

◂ Two polishers working on swords. Note the swords and spears leaning against the wall behind them and the woman who avoids stepping into the sacred workshop area while passing over two swords. A detail from a painting of twenty-five different craftsmen on a pair of six-panel folding screens. (Shokunin zukushi-e, Edo period, Kita-in temple, Saitama Prefecture)

surface and was accompanied by a document stating that it had been placed there in 1274. Scientific examination of the blade indicated that it had apparently been polished with an uchigumori-like stone since the details of the hamon and jigane were clearly visible.

By the Azuchi-Momoyama period (1573–1600) three families—the Hon'ami, Takeya, and Kiya—had become professional polishers and appraisers of swords. The first generation of the Hon'ami family may have served the first Ashikaga shogun, Takauji (1305–58), and the ninth-generation Hon'ami Mitsunori (1554–1619) served ruler Toyotomi Hideyoshi (1537–98) as an official appraiser and judge of swords. In 1594, Mitsunori wrote *Mitsunori katana-ezu*, in which he recorded many details of swords he had seen, providing a clear indication of the quality of the polishing. After the death of Hideyoshi, the Hon'ami continued to be very influential official appraisers under the Tokugawa shoguns until the end of the shogunate in 1868.

During the Keicho era (1595–1610), at the beginning of the Tokugawa shogunate, a new style of sword called shinto appeared. These new swords were practical yet had high artistic value. It seems likely that good polishing was necessary to bring out properties of artistic merit in the shinto swords which frequently had large, wide, and complex hamon with new patterns. Again, however, well-preserved polishes from this period are nonexistent. Be that as it may, other evidence indicates ongoing improvements in the field: the swordsmith Suishinshi Masahide, writing in the late eighteenth century, suggests that the sashikomi nugui finishing technique (which is used to bring out the hamon in strong contrast to the body of the sword, as well as produce a dramatic highlighting of the hamon) could date back to the Momoyama period in the late sixteenth century or the early seventeenth century, at the time of the emergence of the shinto swordmaking tradition.

The Meiji Restoration in 1868 resulted in the restoration of the power of the emperor and the end of the military rule of the shogunate. In 1876, it became illegal to wear swords in public, which led to a drastic decrease in the demand for new swords. Swords became objects valued for their history and aesthetic qualities. With more emphasis on appearance, the state of their polish became increasingly important and techniques improved greatly from the mid-Meiji period. The use of the hadori stone—which historical writings indicate was in use by the early nineteenth century, to create a white, highly, contrastive hamon (*kesho-togi*)—became more common, possibly because this made the swords easier to appreciate for novice Japanese collectors and foreigners alike. It seems likely that, up to this time, the sashikomi technique was more common.

The Sword-Polishing Families

There are several families which were closely involved with polishing over the centuries. The Hon'ami, Kiya, and Takeya families were well known and respected as polishers and appraisers from the end of the Momoyama period and the beginning of the Edo period. All three of these families worked under the shogunate and had high social status. The Hon'ami family is still active in polishing and appraisal today, and the current head of the family is Hon'ami Michihiro. The Hon'ami family can trace its history back to the Muromachi period. In modern times, the Fujishiro family has become very prominent in polishing and appraisal, and the first member of this family was Fujishiro Fukutaro (1878–1949), who worked early in the twentieth century. The current head of the Fujishiro family is Fujishiro Okisato (b. 1947).

The Hon'ami Family

The exact date the Hon'ami family began to work as polishers is uncertain, although it has been suggested that it might have been as early as the Kamakura period. The oldest documented member of the family, Myohon, worked under the first Ashikaga shogun, Ashikaga Takauji. Apparently, Myohon followed Takauji when the leader relocated the shogunate headquarters from Kamakura to Kyoto. By the time of the third Ashikaga shogun, Yoshimitsu (1358–1408), the Hon'ami family had received a title or appointment as *doshoku*. Doshoku were the equivalent of managers or chiefs of small departments with specific responsibilities under the shogunate. These responsibilities could be very specific, such as being responsible for managing tea ceremonies, garden design, flower arranging, or sword polishing and appraisal. The people who were appointed doshoku were allowed to add the suffix "*ami*" to their names, which gave rise to the Hon'ami name. During the Muromachi period many famous artists were recognized in this manner and some of these are Noami for tea ceremony, Kan'ami and Zeami for Noh theater, Taiami for flower arrangement (ikebana), and Zen'ami for garden design.

During the regent Toyotomi Hideyoshi's reign (1585–98), swords became more important, and Hideyoshi created a specific government bureau to be responsible for sword-related matters. This new bureau, the Token Kiwamedokoro, became the responsibility of the ninth-generation Hon'ami, Mitsutoku. This meant that only the Hon'ami had permission to write officially recognized sword appraisals, or *origami*, stamping their documents with an official seal, or *hanko*, provided by Hideyoshi.

Although none of Mitsutoku's appraisal papers remain, evidence that the ninth-generation polisher was a very good appraiser has been gleaned from the book *Katana-ezu*, an illustrated book of swords, and from gold inlays on sword tangs, which were intended to attribute unsigned swords to specific swordsmiths.

During the Edo period, some polishers were officially recognized by the shogunate and received a stipend from the government. There are records from 1792 showing that the Hon'ami family, as well as the Takeya and Kiya families, received payments from the shogunate. The Hon'ami were also responsible for writing certificates of appraisal for specific swords, and initially were the only officially recognized people allowed to do so. On the third day of every month, all eleven branches of the family assembled and examined swords and wrote the certificates of appraisal as appropriate. These certificates were dated, and guaranteed that a sword and its signature or attribution were correct, using the term *shoshin* for these swords. The oldest origami existing today date from 1622 at the time of the tenth-generation Hon'ami, Mitsumuro (1583–1625). There are origami from the tenth- to the thirteenth-generation Hon'ami called *ko-origami* (old origami) that are very well respected. The Toyotomi and Tokugawa governments protected the Hon'ami and initially gave them an exclusive right to pen origami. However, at a later date others were also allowed to write such origami. Official origami could only be written by the main line of the Hon'ami family, and certificates written by branch Hon'ami families or individual Hon'ami family members were called *soejo.*

The Hon'ami family produced a number of famous artisans as well as sword polishers. The most famous was Koetsu (1558–1637), who was renowned for his paintings, ceramics, and calligraphy and was given land by the shogun Tokugawa Ieyasu. Koetsu began his career as a sword polisher, and it appears that the sense of aesthetics, which began with swords, spread to other fields.

In 1607, the shogunate gave the Hon'ami five parcels of land in the Kyoto area with quarries that produced good-quality polishing stones.

The main Hon'ami family distributed the highest quality stones among themselves and its branch families and this helped them to maintain the high standard of their polishing work.

During the Edo period, the Hon'ami family became the head of the shogunate bureau that regulated sword matters, the Katana Wakizashi Mekiki, and were also appointed as *ie-togi*, or polishers of the shogun's swords. Thus, the Hon'ami were very clearly at the pinnacle of the sword world during the Edo period.

THE FUJISHIRO FAMILY

Unlike the Hon'ami family, the Fujishiro family began working as polishers in the early twentieth century, the founding member of this family, Fukutaro, plying his trade sometime before World War II. Fujishiro Okisato is the third generation in his family to pursue polishing and his son Tatsuya (b. 1979), is now the fourth-generation Fujishiro to work as a polisher.

Fujishiro Fukutaro was trained in the Yoshikawa tradition in the early twentieth century, in the late Meiji and Taisho periods (approximately 1900 to 1926). Fukutaro had three sons, Yoshio, Takeo, and Matsuo, all of whom were involved with sword polishing and appraisal.

Matsuo (1914–2004) began to work with his father when he was 14 years old; he left junior high school in his second year to become a full-time polisher. Fukutaro's oldest son, Yoshio (1902–1945), who was 12 years older than Matsuo, also started working with his father at the age of 14, but eventually opened a sword business and became more involved with appraisal than with polishing. Yoshio and Matsuo wrote a sword reference book called *Nihon Toko Jiten* (Who's Who in Japanese Swords; see page 149 for photographs), which became a standard reference book for those interested in the study of Japanese swords, and remains one of the most important reference works today.

According to Matsuo, his father worked swiftly, partly because he had a talent for interpreting the stone markings, which allowed him to proceed with assurance. In general, polishers worked faster in those days than they do today and they worked to different standards. Even so, his father's efficiency allowed him to charge between 16 and 20 yen for a polish where others charged 50.

Matsuo married and moved into his own home-cum-shop in 1937 and became fully independent in 1943. When he began to take on apprentices, he took the more freeform approach that worked well during his training. He believed that his own strengths as a polisher developed because he was not forced to follow a rigid regime such as the systematic approach followed by other schools that lays down precise methods for every aspect of the polishing process. Matsuo encourages his students to find the methods that worked best for them. This means that the various Fujishiro school polishers do polish somewhat differently from each other and are encouraged to use whatever style or technique suits them best. Matsuo's father and his brother, Yoshio, brought him many polishing commissions, and both gave him suggestions and criticisms but left him with all the necessary freedom to experiment and learn on his own.

One particular incident, early in his career, helped to establish Matsuo's reputation. A collector had a Kamakura-period sword by Rai Kunitoshi polished by one of the leading polishers at the time and had paid 50 yen for the polish. However, the polish was not satisfactory and the nioi line defining the hamon was not visible. Yoshio saw this sword and brought it to Matsuo to be repolished. Matsuo repolished the blade and was able to bring out the nioi and a full hamon and charged 20 yen for his work. This occurred before World War II, when Matsuo was 23 years old, and helped cement his reputation. From this time forward, he began to receive many good swords on which to work.

When the war ended in 1945, there was no polishing work available for about two years. Most of Japan was recovering from the war and there was a shortage of food and other necessities. The immediate postwar years were about survival, so there was no work for polishers. In addition, all weapons were confiscated, including traditional Japanese swords, so there were no swords to polish. This resulted in a two-year break in Matsuo's career and that of most other polishers. Matsuo and his family moved away from Tokyo during this period and lived with relatives. However, in May 1946, it was again legal to own swords, and activity resumed in the sword community.

Matsuo and his family returned to Tokyo around the end of 1948. There were apparently only two polishers who were able to make a living working in Tokyo at that time: Matsuo and Ono Kokei, who was a student of Hon'ami Koson, but even they polished mostly for museums. Other polishers found work as appraisers or sword dealers. Matsuo's father returned to Tokyo in 1947 and passed away in 1949.

Around 1964, Japan's economy began to expand rapidly, and this meant that sword-related work also grew. With the increasing amount of work available and a larger number of polishers working, cooperation and competition among polishers has had a very positive effect. Today, there are numerous sword polishers working all over Japan and the quality of the average polish has improved. The average polisher today is clearly better and more effective at bringing out the features and character of a sword than ever before.

STONES FOR FOUNDATION POLISHING

STONE	FEATURE, APPLICATION
ARATO	coarse, used for new swords, or old swords in poor condition
KONGOTO	coarse, used for new swords, or old swords in poor condition
BINSUI	coarse, used for new swords, or old swords in poor condition
KAISEI	coarse, used in diagonal direction; lengthwise direction for natural stone
CHU-NAGURA	used in lengthwise direction
KOMA-NAGURA	used in lengthwise direction
UCHIGUMORI HATO	used in lengthwise direction
UCHIGUMORI JI-TO	used in lengthwise direction

STAGES FOR FINISH POLISHING

STAGE	FEATURE
JIZUYA (NARUTAKIDO STONE)	a piece of thin stone is used to go over the ji to bring out the jihada; jizuya refers both to the process and the polishing device made from the narutakido stone
HADORI (HAZUYA: UCHIGOMORI STONE)	a piece of thin stone is used to go over the ha of the whole blade, including the boshi
KANAHADA NUGUI (IRON OXIDE IN OIL)	the preparation is used to go over the ji, shinogiji, and mune to make them darker so that the ha stands out against the ji
SASHIKOMI NUGUI (MAGNETITE IN OIL)	the preparation is used to go over the entire body of a blade except the kissaki; this method is used as an alternative to kanahada nugui
MIGAKI (BURNISHING SPATULA, BURNISHING NEEDLE)	used to produce mirrorlike finish on mune and shinogiji
YOKOTE SUJIKIRI (HAZUYA)	used to make the yokote line
NARUME (HAZUYA)	used to polish the point area

A NOTE ABOUT USING STONES: The polishing stones used up to the chu-nagura stage are applied in a diagonal direction relative to the sword's lengthwise axis. The earliest stones (arato and kongoto) can be used at an angle perpendicular to the sword's lengthwise axis. The fact that each successive stone is used at a slightly different and shallower angle relative to the lengthwise axis makes it easier to recognize and polish over scratches made by the previous stone. However, from the chu-nagura stage, all of the polishing stones are used in a lengthwise direction parallel to the sword's lengthwise axis. The general sequence of stones used in polishing a sword is given in the charts above.

The polisher should begin with the finest stone he can use for a particular sword, since rougher stones will remove more metal, wearing down the sword. When polishing new swords—which are in a very rough condition, need their shapes refined and finished, and require sharpened and formed edges—the polisher must begin with arato or kongoto stones and work his way up to the finer stones. For severely rusted and damaged swords, the same approach is taken. In the case of an older sword that is lightly rusted but otherwise undamaged, the polisher might be able to begin with the chu-nagura or koma-nagura stone, which will prevent wear on the sword and greatly shorten the polishing time.

Part II

The Polishing Process

Chapter 5 BACKGROUND

Part Two describes the Japanese sword-polishing process in sufficient detail to enable the reader to develop a thorough grounding in the craft. The text and photographs document the work of one polisher, Takaiwa Setsuo. The principles and techniques depicted apply to most Japanese sword polishers, although variations in techniques, tools, and materials do exist since some polishers tailor them to suit their abilities and needs. However, the basics as well as the primary materials used provide a common foundation for all sword polishers.

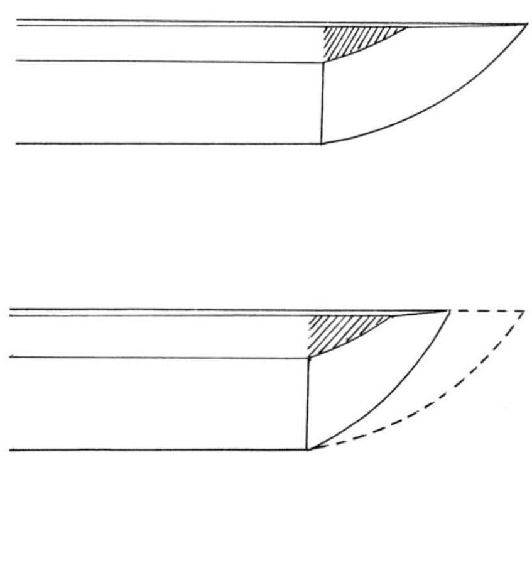

A well-balanced kissaki (*top*) and an unsightly kissaki (*bottom*). The ko-shinogi line must be parallel or complementary to the shape of the kissaki. The line should parallel that of the kissaki's cutting edge. The dotted line shows the original edge of the kissaki. It has been polished off and the shape of the ko-shinogi has not been adjusted.

Even assuming that all polishers use basically the same methods and tools, there is a wide range in the results which are affected by three major factors:

1. Experience. Some polishers with twenty or thirty years' experience will have improved or modified their techniques while others with maybe only a handful of years under their belt will be less self assured.

2. The swords. Not all swords are the same. The steel is handmade for each, and the iron ore is obtained from different sites and refined in different places by different people. In addition, swordmaking methods have changed over the centuries, which means that steel from different eras and localities will have different compositions, hardness, appearances, hamons, and so on. All of these properties affect the steel, and how it responds to the polishing stones. Moreover, modern swords respond differently to a stone than, say, an eight-hundred-year-old specimen.

3. The polisher's judgment. A polisher must understand swords thoroughly to perform at his peak. A large part of the final outcome in the appearance of the sword will depend on the polisher's vision of what the final sword should look like. With sufficient skill a polisher can draw out the sword's most impressive details. But numerous judgments are called for. For example, the jihada can be strongly emphasized, but this might make utsuri (if it is present) difficult to see. If the blade is a Koto Bizen sword (see Appendices, page 169), the utsuri (if present) should be brought out since it is often a hallmark of that school. Thus, a large part of the polisher's work is to decide how to polish a particular sword and what parts should be enhanced or emphasized, such as the jihada, hamon, utsuri, or other aspects of the blade. Unfortunately, emphasizing one of these properties could leave others less clear or nearly invisible. Thus, a good polisher not only has a good command of his techniques, but can evaluate a sword in order to realize its full potential.

On the Shaping of the Sword

While the swordsmith makes the sword, it is the polisher who actually finishes it, making sure that its fine and subtle details are correct.

If a sword is to look good and have the correct proportions, it has three main areas that need particularly careful shaping, since they are often worn down by use or polishing. If they are carefully shaped and formed by the polisher, the final appearance of the sword will be aesthetically satisfying. The areas are the *machi* at the beginning of the sharpened part of the sword; the cutting edge where the yokote meets the edge; and the *ko-shinogi* (the part of the shinogi over the point).

The machi is the base of the blade above the tang, where the cutting edge begins. The area comprises the munemachi and the *hamachi*, which marks the beginning of the sharpened edge. The lines of the mune and *ha* (cutting edge) must follow the shape of the blade and not taper or flare excessively. Although the machi is usually the widest part of the blade, it must nevertheless be in proportion to the rest of the blade (see pages 171 and 172).

The shape of the area where the yokote line meets the cutting edge is also important, for it marks the start of the kissaki. The curves here should form a defined, obtuse angle at the yokote. The polisher must take care not to polish or wear away too much of the ha at the point where the yokote meets the edge, since this would result in a poorly formed kissaki.

Finally, the ko-shinogi, which defines the upper boundary of the kissaki, must complement the shape of the kissaki. This is achieved by making the ko-shinogi line parallel the edge of the kissaki.

A Cautionary Note for Amateur Polishers

The major danger to a sword during polishing is that too much steel might be removed, thus forever altering or ruining the shape. This can occur at the earliest stages of the polishing process, from the first stones up to the koma-nagura stage. The stones used after the koma-nagura are very fine and not likely to alter the shape of the sword. Thus, if sword owners would like to attempt to polish a sword themselves, it is advisable to begin with a stone, such as suita or uchigumori, that is fine enough to minimize the risk of permanently altering or damaging a sword. Although finer stones cannot remove deep rust, chips, or damage to the point area, or begin to polish newly made swords, they are unlikely to increase existing damage or create new damage. Ideally, nonprofessionals will be able to find advice or guidance before attempting to restore a sword, or will gain experience with the stones at, or finer than, the suita-level stones before moving on to use coarser stones.

The Purpose of Polishing

Gendaito, or modern swords, are intended to be appreciated as art, but must also be functional as a weapon. Because gendaito are no longer simple weapons, much more care is given to highlighting their beauty; in fact, much more effort must be invested in these swords than strict utility would require. As a result, the role of polishing them is now far more important than in the past, because polishing is the final step which makes all the intricate and individual details of the sword apparent to people who want to observe and appreciate them. A good analogy for this situation is to consider the sword to be a precious gem which must be cut and polished before its real beauty can be seen and appreciated. Thus, the steps or requirements of the polishing process necessary to finish and allow appreciation of a sword can be listed as: (1) a sword needs to have a well-balanced shape for aesthetic appreciation as well as for practical reasons; (2) the inherent features of the sword and its steel must be brought out and made visible, and these features include the details of the jigane (surface) such as the jihada, chikei, utsuri, ji-nie, *yubashiri*.

Similarly the details of the hamon must also be made visible and these include the nie, nioi, *sunagashi*, ashi, and kinsuji. The polisher's role is to enhance the value and beauty of a sword by making all of its critical features easily visible. In addition, it should be kept in mind that no part of a sword, except for the mune and shinogiji, has flat surfaces. The remaining surfaces (in fact, most of a blade's surfaces) are curves and arcs. All the complex features—the steel's surface and the shape—mean that the level of skill and the technique demanded of a sword polisher are extremely advanced.

In addition to polishing and finishing newly made swords, the polisher must restore and repair old swords. Thousands of valuable older swords are in museums, galleries, and private collections, and these blades must be maintained. There is a Japanese proverb related to the sword and rust: "It is better to wear out than rust away." Polishers, however, are required not to wear out a sword while removing rust and damage in the process of restoring an old sword.

The polishing area and setup. Note the plastic sheet polisher Takaiwa Setsuo uses in the early stages of foundation polishing.

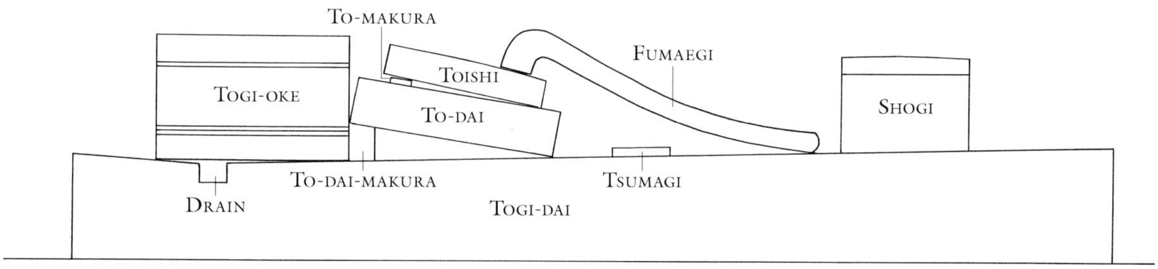

Polishing area with standard tools.

The Polisher's Workshop

In setting up a polishing workshop, it is advisable to select a space that can be devoted to polishing and nothing else. A clean, uncluttered location is desirable; it must be quiet enough for the polisher to concentrate on his work without distraction.

It is also a good idea to set up two work areas, one area for foundation polishing and a second for finish polishing. The reason for this is that the early-stage stones are very coarse, and any particles from these stones present during the finishing stages can leave deep scratches on an almost finished blade. It is hard to remove all traces of the coarser stones, especially the synthetic ones. Takaiwa has to use one area for all of his polishing. He solves this problem by covering most of the work area with a plastic sheet during the early stages of foundation polishing, and removing it to clean thoroughly before proceeding to the finish polishing.

Dust

One of the biggest enemies of a good polish is dust. If airborne or surface dust falls between one of the finishing stones and the steel, the polisher will inadvertently put deep scratches into the surface. Thus, a dust-free environment is vital for a polishing shop. Locating a work space away from areas with a lot of automobile traffic and districts where other types of work are done is vital to limiting the amount of dust.

The Workshop's Size

A polisher's work area should be neither too big nor too small. A clean, quiet area about 10 by 10 feet (3 x 3 m) would provide a comfortable area in which one person could work.

Lighting

The quality of the lighting is a vital component of a polisher's work area. From ancient times, it has been said that the windows of the workshop should face north and never south. The direct light from a southern exposure will change in brightness and intensity during the course of the day. A polisher needs a uniform and consistent light in which to work, especially during the finish polishing stages. That means that there should be a good lighting source in the shop that can provide a consistent quality of illumination throughout the day, from morning to evening.

AIR CONDITIONING

Water is necessary to lubricate the stones as well as keep the area clean. However, this raises the humidity in the room. As blades can easily rust, they can be protected from the humidity by a good air-conditioning system.

The Polisher's Tools

Polishers work on a *togidai*, which is a raised wooden platform that serves as a compact work space about 5 by 3 feet (1.6 by 1 m) in size. Water is constantly used during the polishing, so the togidai must drain well. To provide drainage, the floor of the togidai usually slopes about 3% to lead the water away from the polisher and the work area. Generally, the floor is made of a softwood, such as cypress or cedar, which is favored because it dries easily. The polisher's tools and implements such as the water bucket, work block, clamp, footrest, and stool are made of softwood in order to protect the blades.

TOGI-OKE: WATER BUCKET

The bucket used by polishers is usually 18 by 14 by 7 inches deep (45 by 35 by 18 cm). Water from the bucket is used constantly during the polishing process to help lubricate the blade and the stone. The polisher also uses the water constantly to cleanse the surface of the blade in order to examine the progress of his polishing. A water bucket usually sits over the drain, in front of the polisher.

The bucket should not be employed outside the polishing room. Having a dedicated water bucket ensures it is not exposed to harmful agents such as acids or other materials that might later come into direct contact with the blade, resulting in rust or other problems.

Sometimes washing soda or sodium carbonate is added to the water. Washing soda helps the blade slide smoothly over the stones and, because it is alkaline, it protects the blade from rusting and from the acidity of sweat and the oil on fingers. However, the synthetic stones used for the first polishing stages are easily damaged by washing soda. Since synthetic stones kept for a long period in water containing sodium carbonate begin to deteriorate and become brittle, it is probably a good idea to use water without sodium carbonate, or to dissolve only a small amount in the water used with synthetic stones (one teaspoon for one bucket of water).

To-dai: Wooden Work Block

The to-dai is a wooden block used as a base or rest for the polishing stones. Usually made of cypress, it is an average size of 3½ (height) by 14 by 5 inches (9 by 35 by 12 cm). Polishers will cut the block to a length which they find comfortable, and position the stone at a convenient distance between the stool and the water bucket.

To-dai-makura: Small Wedge

A piece of wood is wedged under one end of the stone to adjust its angle on the wood surface. The location of the to-makura varies, depending on the polisher's needs. It is approximately 3 by ³⁄₁₀ by ¹⁄₁₀ inches (7 by 0.7 by 0.3 cm) in size.

To-makura: Resting Block

The purpose of the resting block is to raise the end of the polishing stone so that it is conveniently angled for the polisher, enabling him to work in an effective and comfortable position. One of the edges of the resting block has a notch to hold the work block in place. The front of the work block fits into the notch and the to-dai-makura rests up against the bucket. The wood used is cypress and the typical dimensions of the resting block are 4 by 4 by 6 inches (10 by 10 by 15 cm).

Fumaegi: Clamp

The fumaegi is a unique tool. It is a curved piece of wood that serves as a clamp to hold the polishing stone in place. It provides adequate pressure to hold the stone immobile, spreading the pressure over a large area of the stone so that one can easily reposition or change a stone. Since the stones are delicate, a high-pressure clamp can split or crack them. The fumaegi has a hooklike formation at one end which fits on the top of the stone, while the other end serves as a platform for the polisher's foot, with which he applies pressure to hold the stone stationary.

Polishers make their own fumaegi, so these unusual clamps vary in shape and length to accommodate polishers' needs.

Tsumagi: Footrest

There are two reasons for using the footrest. First, the toe or ball of the polisher's right foot is placed on the footrest for stability and comfort while the weight of the polisher's right heel presses down on the fumaegi. Second, the footrest keeps the polisher's right foot off the to-dai surface, which is wet most of the time because water is constantly being used. The size of the footrest is 1 (height) by 4 by 4 inches (2 by 10 by 10 cm). The wood used is usually Japanese cedar.

Shogi: Stool

The polisher sits on a small stool while polishing. The size of the stool is 5 (height) by 16 by 12 inches (12 by 40 by 30 cm) and it is usually used with a cushion or pillow.

Saide: Rag

The polisher wraps a cotton cloth tightly around the blade in order to provide a place by which he can grip it. The cloth needs to be clean and dust-free.

Togizuka: Extension or Grip

When polishing a small blade, the polisher often uses an extension to give him a better grip on the blade. The tang of the small blade fits into the togizuka, allowing the polisher to grip the blade while working. To avoid damaging the tang, the polisher needs to be careful that it fits properly and securely into the extension. Some polishers prefer not to use an extension.

Takaiwa examines the condition of a sword before he begins polishing.

Chapter 6 FOUNDATION POLISHING

Polishing the Different Parts of the Sword

The following descriptions apply both to new swords and those that are severely rusted or damaged. It is very important to use the proper techniques and methods from the first step (the arato or kongoto stone) if the final polish is to be successful. If any of the stones are used improperly or insufficiently, the resulting errors or undesirable marks cannot be corrected with subsequently finer stones, so it is important to use a stone until the proper level of finish is obtained.

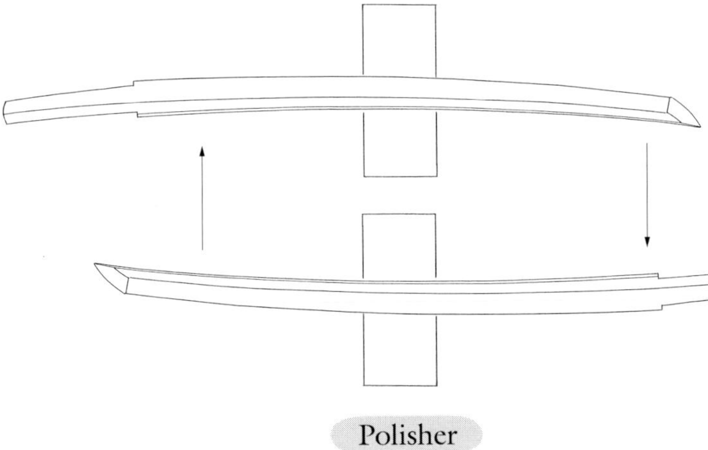

When polisher Takaiwa Setsuo has polished two-thirds of the length of the sword, he reverses the direction in which the sword points, so that he can polish the remaining third (closest to the tip) comfortably.

In general, a polisher positions the sword in a manner that works best for him, whether working on the mune, shinogiji, ji, or ha. Taking into consideration comfort and results, he may alter the angle of the sword being polished, the way the cutting edge faces (toward or away from the polisher), and so on. The various schools of polishing have many detailed rules concerning positioning the sword, but an individual polisher will often alter them.

Sitting Position

The polisher assumes a traditional sitting position. It appears awkward, low, and difficult, but it does allow the craftsman to position himself close to his work.

The polisher sits on a low stool and leans forward over the sword. His right heel presses down on the clamp to hold the stone in place, while his right shoulder hovers over his right knee and his head hangs over the section of the stone he is using. While this position is both practical and effective, it can take a while for one to become accustomed to maintaining it for a lengthy period of time.

He lowers himself onto the stool without moving his feet.

Takaiwa stands in the proper position before sitting, one foot on the fumaegi, the other on the low, cushioned stool.

Resting on his heel, almost in position.

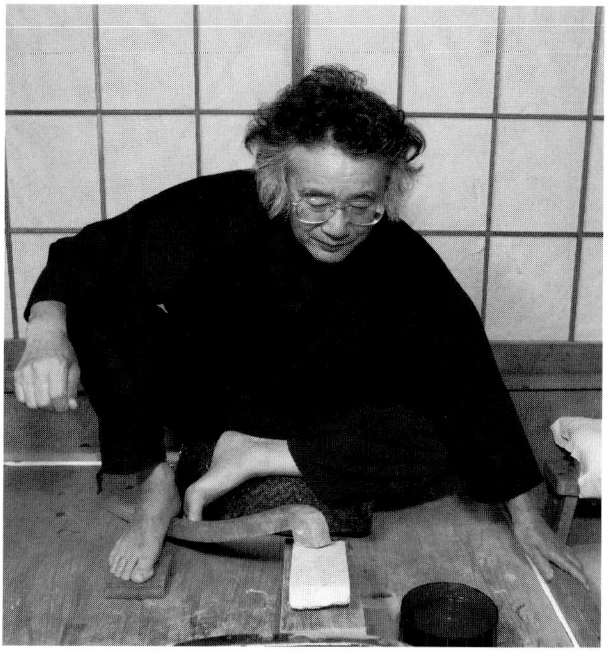

He sits on the stool and slides his left leg forward until his left foot touches the fumaegi.

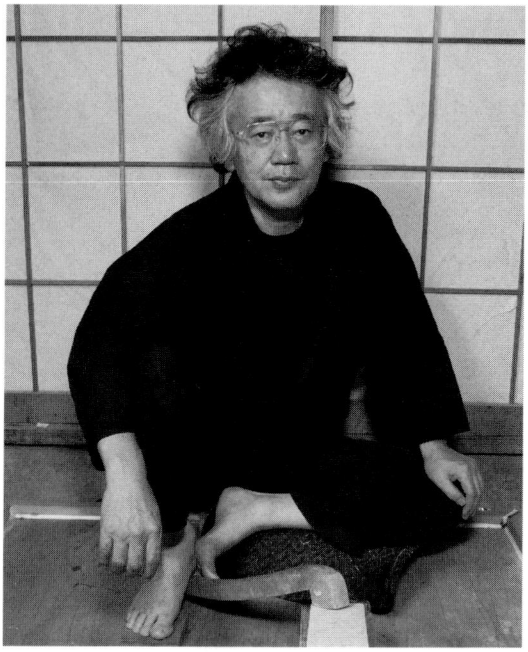

In this position, he is seated, with all his weight off of his feet.

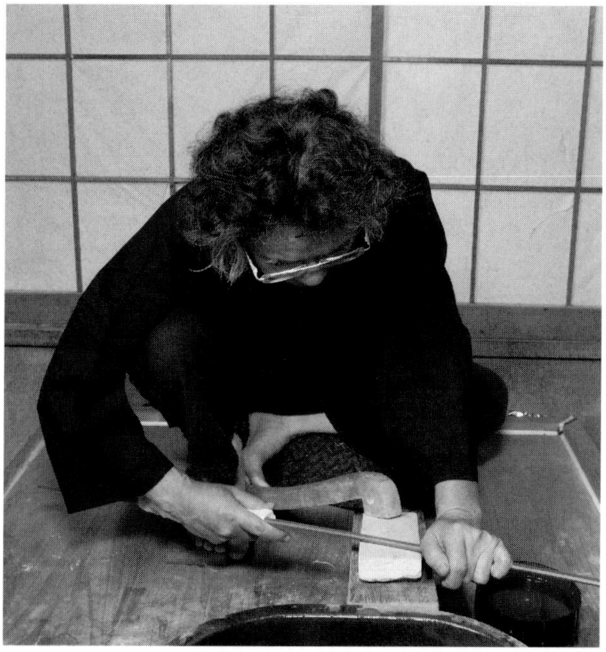

To polish, Takaiwa leans forward to grasp a sword. His right shoulder is over his right knee and his head and shoulders are over the sword, giving him optimal control of the sword and stones.

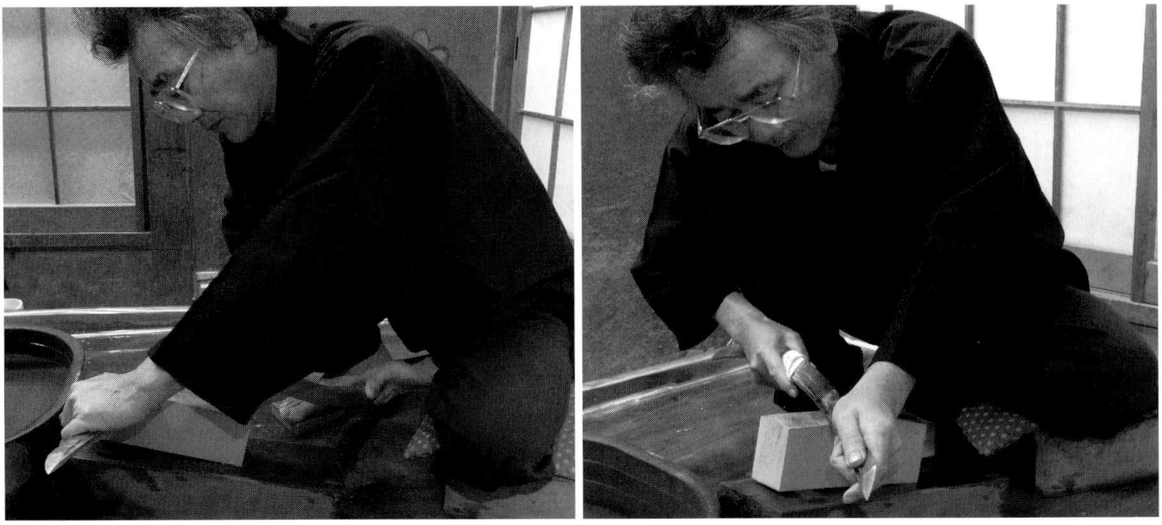

These photographs show the polisher in the proper position. In both instances, his head and shoulders are almost directly over the part of the blade being worked on, and he can apply pressure to the blade while keeping a close eye on the process. Note that his eyes are directly over the portion of the blade on which he is working.

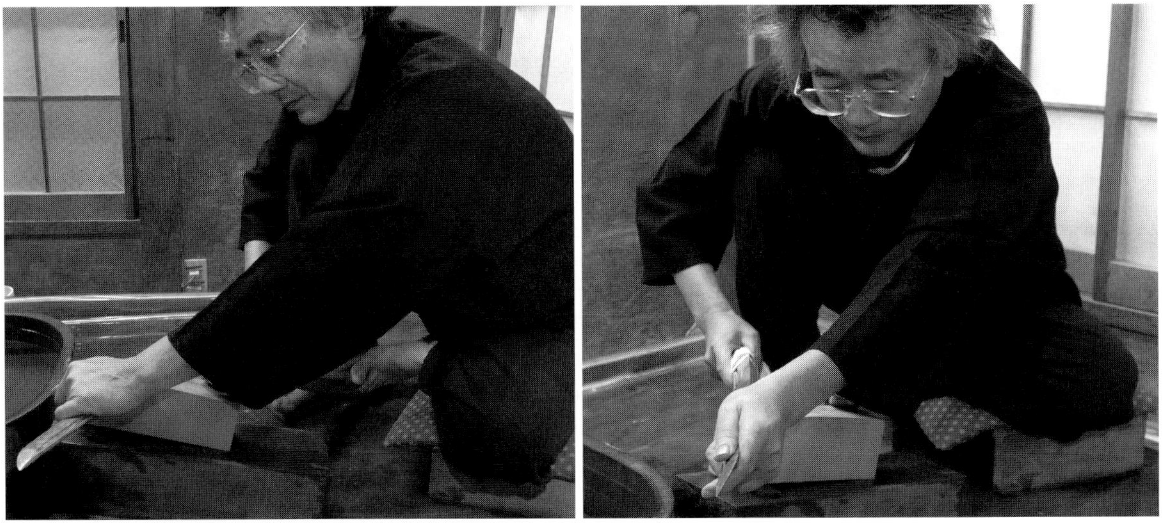

Two examples of bad positioning. The polisher is positioned too far back. His eyes and shoulders are not over the blade, making it difficult to apply the proper pressure on the blade.

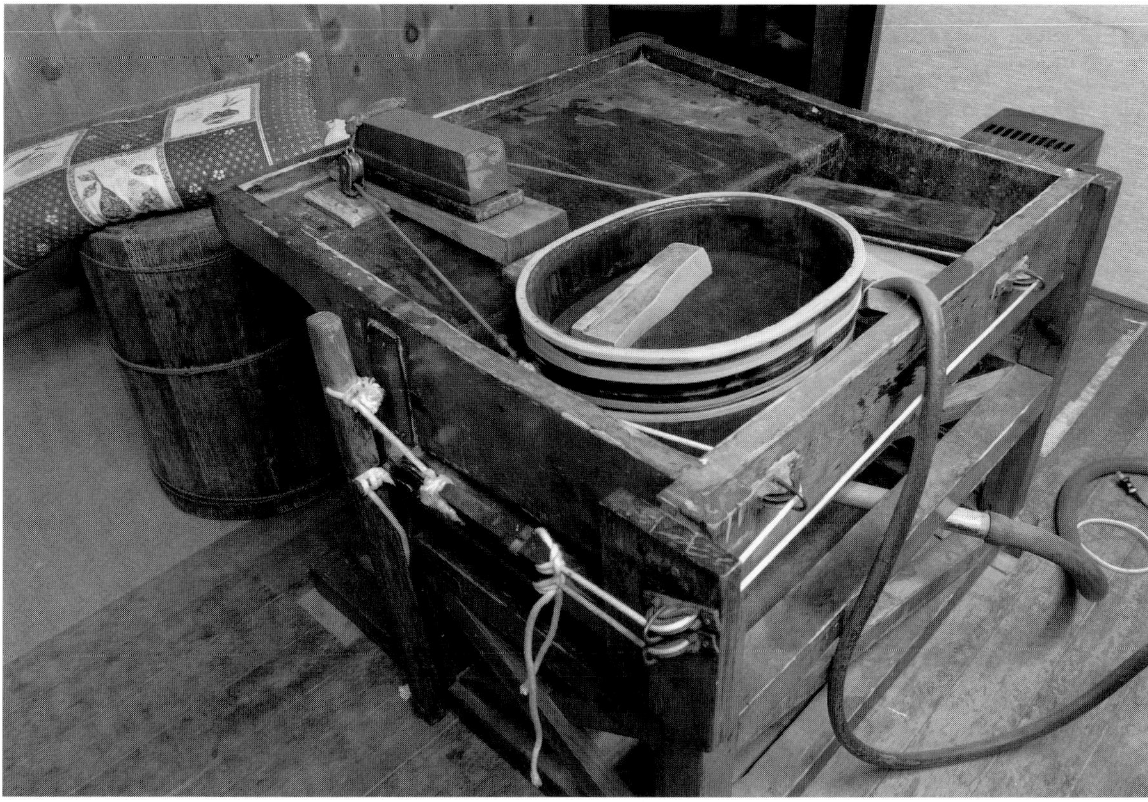

The work setup, preferred by polisher Ogasawara Akira allows him to polish swords comfortably.

The wooden work block and the polishing stone are firmly fastened with wire because a foot-held clamp (fumaegi) cannot be used here

Straightening a Sword

Often a sword is bent or twisted from the heat stresses produced when the hamon is made, or simply from use or abuse. The blade must be straightened before the polisher can proceed. There are several methods to repair warped blades.

The blade can be straightened with a pair of *tamegi*. This is a set of two thick wooden blocks with slots in them. The sword is set into the slots, back-first, with the distorted section of blade centered between the two bars. Next, boiling water is poured over the blade at the bend. Then, the polisher applies leverage to straighten the blade by pulling or pushing the handles of the tamegi toward or away from each other. The pressure must be on the shinogiji and not on the cutting edge, which can be chipped or cracked with too much pressure. Athough their use appears simple, it is probably advisable to discuss using tamegi with someone experienced before using them yourself.

A blade can also be straightened with a hammer and a wooden block that has a slightly concave surface. The sword is placed on the block and raised, and the bent section is hammered down. It takes experience to perform this operation successfully. Often, hammering one part of the blade introduces some warping in another part, which must then be corrected. In addition, the hammer must only strike the blade on the shinogiji, and not on the ji or on the ha, which can be permanently damaged.

The sword can also be straightened with a smaller hammer called a *hazuchi* and an iron block. This method is often used to straighten the

A sword is being straightened using two heavy wooden blocks with long handles to provide leverage. This method often allows bends and small curves in the blade to be corrected. The technique is usually used for new swords or swords that have bent with use.

A heavy hammer is being used here to correct a severe bend in a thick blade. The hammer blows are focused on the shinogiji area. The blade is resting on a wooden block with a concave surface.

Here, a sword is being straightened with a smaller hammer than shown in the previous photograph. However, in this case, the blade is resting on a flat iron block. This arrangement is used to straighten the area near the point.

A sword is being straightened with a U-shaped steel bar and a vise. This method can be used to correct a twisted or bent blade. Notice that a sheet of copper has been placed between the jaws of the vise and inside the U-shaped bar to prevent steel-to-steel contact.

area near the point. The blade must be struck very lightly and on the shinogiji. Badly aimed blows can permanently damage or chip the cutting edge or point.

For twisted swords, a vise lined with a copper sheet and a U-shaped iron bar also lined with copper are used. The blade is straightened slowly by twisting the blade with the bar. The U-shaped bar is slipped under the blade to apply pressure to the shinogiji. This keeps the pressure away from the edge and the hamon to avoid chipping or cracking. The copper prevents any part of the sword from coming into direct contact with the steel vise or bar.

The sword has many surfaces and these are usually polished in a standard sequence. For swords without grooves, the surfaces are polished in the following order: the mune, the shinogiji, the ha, the ji, and the kissaki.

For swords with grooves, the surfaces are polished in the following order: the mune, the shinogiji, the ha, the ji, the grooves (*hi*), and the kissaki.

Mune: Back Edge

OBJECTIVE

To straighten the sharp ridge line along the center of the back. First, it is necessary to check to see if the blade itself is curved, or whether just the ridge is curved or wavy. This first part of the polishing process corrects any waviness or curvature in the ridge. Once the polisher can gauge where the ridge should be, he can begin by polishing the sides of the ridge to correct the deviations in the line of the ridge.

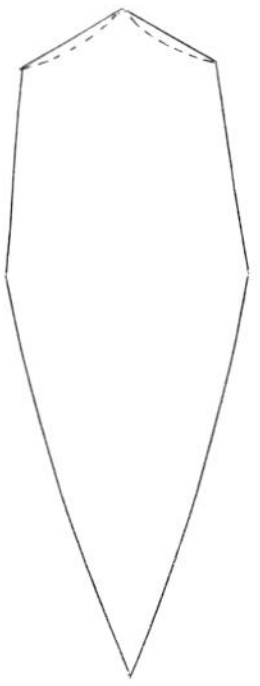

A cross-section of a shinogi-zukuri sword (see page 173). A flat stone should be used to polish the sides of the mune, otherwise it may end up concave rather than flat.

STONE

At this stage it is necessary to use a stone with an absolutely flat, horizontal surface. If the polisher uses a stone with an uneven or rounded surface, the sides of the ridge may become concave, rather than straight.

POLISHING METHOD

Because the mune is polished differently from the rest of the blade, it is best to look at the overall polishing process. Since there is no need to produce the jihada on the mune and the final polishing is done by burnishing (see chapter 7), fewer stones are used to polish it. After polishing with a synthetic nagura stone, sandpaper (grade nos. 1200 to 1500) is used. A piece of sandpaper is wrapped around a small block of wood. A few drops of oil make the process smooth and effective (see groove polishing, page 85). The mune is then burnished with burnishing needles when the shinogiji is burnished.

Polish from the *munemachi* (the begining of the mune at a point above the tang) and toward the kissaki. The sword is moved in a diagonal direction across the stone. The polisher finishes one side of the ridge before beginning the other. He should check the ridge frequently to be

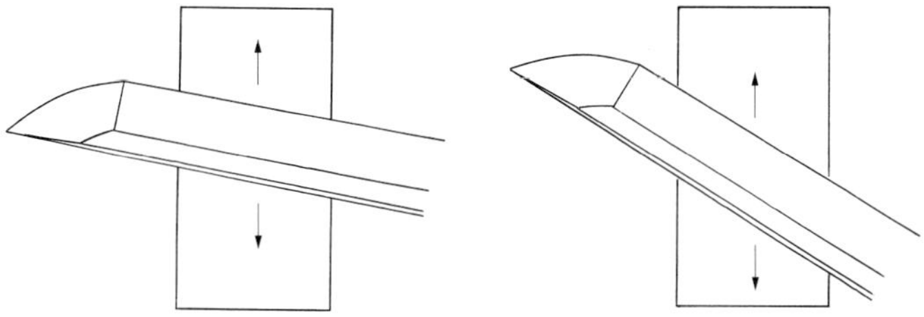

Moving the sword over the stone as shown will produce markings at approximately a 90-degree angle (*left*) and 45-degree angle (*right*).

Takaiwa polishing one side of the mune at the point area. The stone leaves marks at an angle of approximately 30 degrees to the lengthwise axis of the sword. From the chunagura stage on, all the stones used on the mune will leave markings in a lengthwise direction.

Takaiwa polishes the other side of the ridge on the mune. The process is the same as that described in the previous photograph. Notice that the sword edge is tilted away from him, while in the previous photograph the sword's edge is tilted toward him.

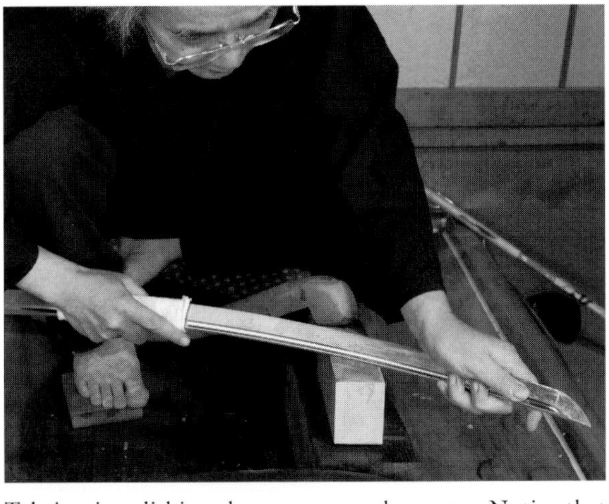

Takaiwa is polishing the mune near the center. Notice that the sword edge is angled toward him.

This is a closer view of Takaiwa polishing the other side of the mune, again at the center. Notice the clamp holding the stone in place and also how the blade is tilted to allow him to work one side of the mune.

sure it remains straight, and that the corrective realignment of the ridge is going as planned. If the polisher is working on a sword in a reasonable condition, he may only have to correct uneven or curved areas. In this case, the polisher can move the sword over the stone in a shallower diagonal stroke and cover a wider area of the blade as he works.

If the blade needs a large amount of work, it is moved diagonally over the rougher stones, perhaps at an angle very close to perpendicular to the lengthwise axis of the blade. If less work is needed and the corrections are minor, the blade can be moved at a shallower angle over the stones relative to lengthwise axis of the blade. In this case, it would be possible to polish a wider (3-inch/8-cm) area of the blade with each stroke.

The force applied to the blade should be in a ratio of 6:4 between the left hand and the right hand. The left hand leads the sword over the stone, while the right hand controls the angle of the sword. This rule applies to all the steps and stones during the entire foundation polishing process. The polisher should polish with the idea of pushing and pulling the sword.

Shinogi and Shinogiji: Ridge Line and Ridge Area

OBJECTIVE

To make the shinogi and shinogiji flat.

STONE

A stone with an absolutely flat surface is essential. A rounded stone might leave the polisher with a concave shinogiji.

POLISHING METHOD

The sword must be moved over the stone in precisely the same direction and angle along the entire length of the shinogiji. If this is not done, the resulting markings on the sword will be present in a range of different angles. File marks in perfect alignment help produce a beautiful, sharp contour along the shinogi line. At this stage, the plane of the shinogiji should be flat, even if the shinogi line is not perfect. Usually one polishes diagonally over the stone (relative to the lengthwise axis of

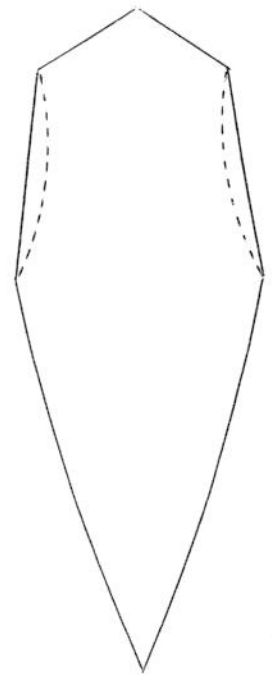

A cross-section of a shinogi-zukuri sword. As in the previous step, the sides of the shinogiji may end up concave rather than flat if the polisher uses an uneven or rounded stone.

the blade; see figure on page 71), covering about 2 ½ to 3 inches (6 or 7 cm) at a time, but if the sword is highly curved, the area covered with each stroke will be less. When polishing with the sword's edge directed away from him, the polisher should use the part of the stone furthest from him; when polishing with the sword's edge facing toward him, the polisher should use the part of the stone closest to him. One side is polished entirely from the munemachi to the kissaki, before the polisher goes on to the other side. In principle, as described in the mune section (page 70), all the polisher needs do for swords that already have a good shape is to correct an uneven surface. In this case, he can polish at a diagonal angle (relative to the blade's lengthwise axis). If a large amount of correction is necessary in different areas of the shinogiji, the polisher must polish small areas at a time, and will have to move the blade over the stone at an angle almost perpendicular relative to its lengthwise axis.

HINTS

At this stage an imperfect shinogi line is acceptable (in other words, the width of the shinogiji does not need to be uniform along the entire length of the sword). It is more important that the plane of the shinogiji is flat. The width of the shinogiji can be adjusted at the next stage when working on the ji. However, the shinogiji around the kissaki should be close to the desired shape. In addition, the polisher need not match the shinogiji and the ji perfectly until later; stopping to check on this now will interfere with the finish and shape of the shinogi line. Sometimes, when a sword has a slight depression in the shinogiji that is polished away, the shinogiji will become wider, and the line of the shinogi will descend into the ji. However, this will be corrected at a later stage when the ji is polished.

Ko-shinogi: Ridge Area at the Point

OBJECTIVE

To make the ko-shinogi absolutely flat. The ko-shinogi is the portion of the shinogi that extends into the kissaki, and in a broader sense the triangular area defined by the lines of the ko-shinogi, kissaki, and mune. This part of the shinogi begins with the same width as the rest of the shinogi, but tapers as it draws toward the tip of the sword, before disappearing. Thus, it is important to define this section carefully to preserve the shape of the point.

STONE

Move the sword using small along the lengthwise axis of the stone, along either its left or right edge.

POLISHING METHOD

The polisher should work from the direction of the tang toward the point, and progress slowly and carefully up to the tip. He must polish in a way that progresses gradually from the yokote area toward the tip of the point. The polisher should move the sword gently and carefully to remove depressions and irregularities in the ko-shinogi surface. The ko-shinogi line should have a gentle curve that parallels the edge of the kissaki.

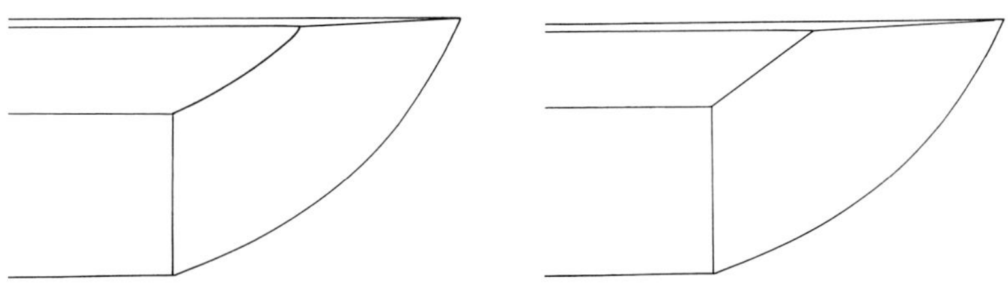

The ko-shinogi line should have a soft curve (*left*) rather than run straight (*right*).

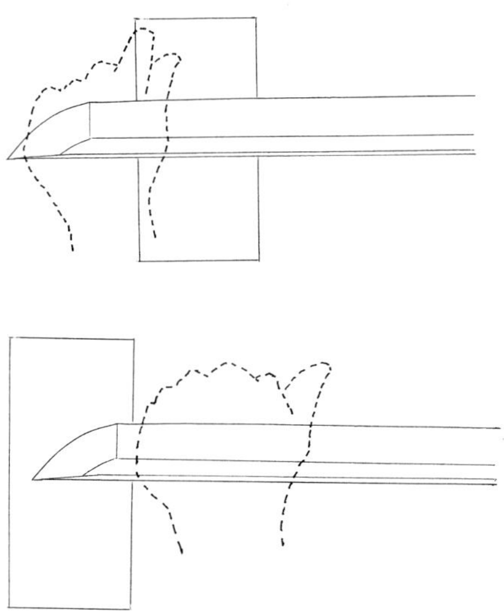

Brace the point area as much as possible. When this becomes impossible, move the sword to the right side of the stone.

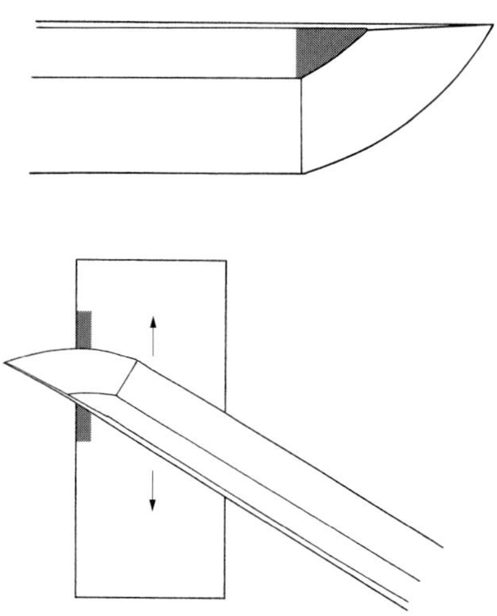

The area near either the left or the right edge of the stone is used when the ko-shinogi is polished.

When polishing the kissaki, you should try to support this small area as much as possible with the left hand for greater stability. When this becomes impossible (i.e., when very nearly all of the blade's point is over the stone), move the sword to the right side of the stone and steady it with the left hand placed alongside the stone, while gripping the sword lower down on the blade, closer to the tang, with the right hand. Either the area near the left edge of the stone or that near the right edge of the stone is used when the ko-shinogi is polished. It depends on which side you are polishing.

Takaiwa polishing the ko-shinogi, using the left edge of the stone for one side and the right edge for the other.

Ha: Cutting Edge

OBJECTIVE

To give the sword a sharp cutting edge and define the edge's overall line. The line of both the ha and mune must be perfectly straight, continuous, and uninterrupted.

STONE

Use a flat or a slightly rounded stone. Takaiwa uses a flat one.

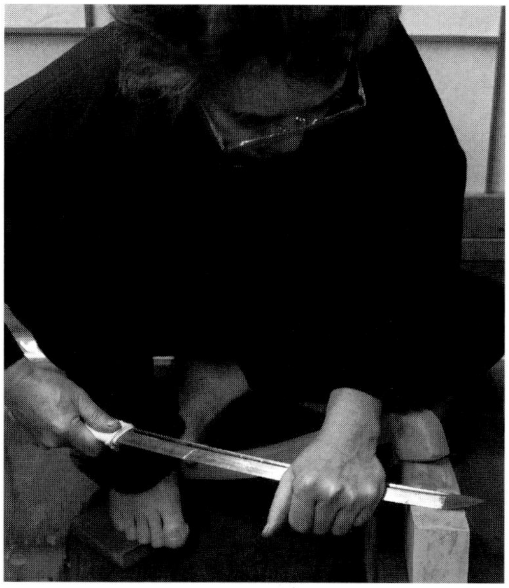

Takaiwa is working on the edge near the yokote, moving the blade along the lengthwise axis of the stone.

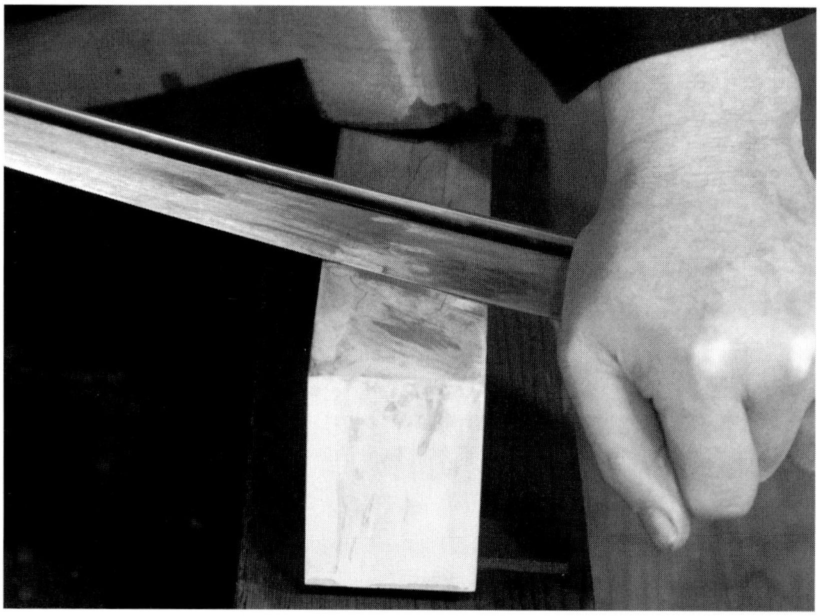

A closer view of Takaiwa polishing the edge. He uses the part of the stone furthest away from him when he polishes the ha and the ji, with the edge facing away from him.

POLISHING METHOD

The cutting edge region is about ⅓ inch (7 to 8 mm) wide and extends from the top of the tang (hamachi) to the yokote line that defines the beginning of the point (area 1 in the figure below). The sword is moved at right angles to the stone and perpendicular to the lengthwise axis of the sword. The area worked on at one time is very small, and the blade is moved up and down a section of stone about 3 inches (7 cm) long. This part of the polishing can take a long time. The edge must be shaped carefully, and because of its hardness, progress is slow. When area 1 is finished, the polisher begins area 2 in the same way. In other words the edge (area 1) is corrected and shaped first, before work begins on area 2 above it. It is important that the appearance of these sections be uniform. Japanese swords are shaped so that the contours of the blade drop all the way down to the cutting edge itself: there is no bevel so that there is an edge just above the sharpened edge.

From this point many polishers work their way up the blade, from areas 6 to 3. Takaiwa prefers to work down from the shinogi line for new and old sword's alike. This method is particularly effective for a worn or damaged blade. After finishing areas 1 and 2, he proceeds immediately to the ji (area 3), a flat rather than curved surface. It is important that the finish, appearance, and direction of the marks left by the stone be the same for each section.

The angle between the shinogiji and the ji is shallow. To help preserve the shinogi and appearance of the sword, the width of the area polished at one time will be wider than in area 1; it is important not to intrude into the shinogiji. In practice, this means that the blade is polished at a shallow angle, beginning at about 45 degrees relative to the lengthwise axis of the sword, and the polishing strokes over the sword are longer.

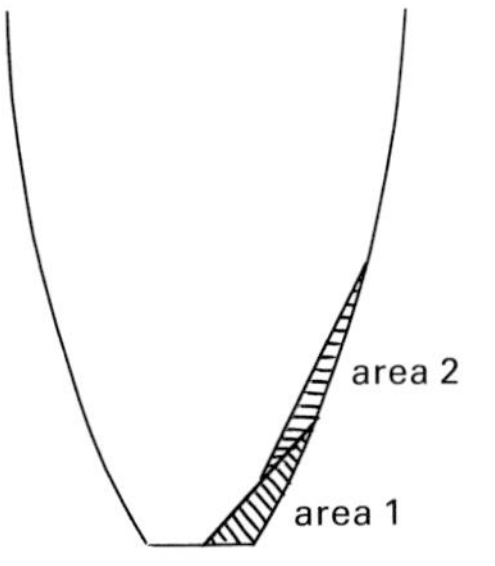

A cross section of the edge of a shinogi-zukuri sword. Area 1 comprises the prime cutting area, which is corrected and shaped first, before work is done on area 2.

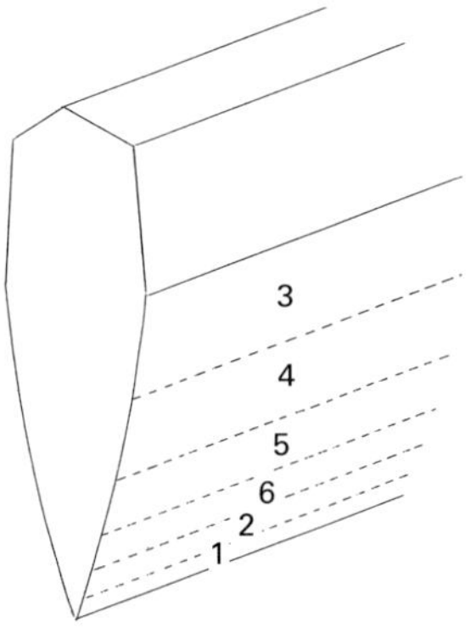

The numbers given on the edge and the side of the blade indicate the order in which Takaiwa works on the areas.

HINTS

It is a good idea to polish the *nioi-guchi* area so that it is flat and of the same thickness or height as the surface around it. Some schools deliberately leave the nioi-guchi thicker than the surrounding area of the blade surface, but if it is too thick the nioi line will appear too sharp. What is desired is a wide diffuse, nioi-guchi line.

In addition, it is important not to excessively polish the yokote area and the hamachi area. It is tempting to overpolish the yokote area in a bid to make it more distinctive, while the hamachi area is easily overpolished just because this area is the heaviest part of the sword and rests strongly on the polishing stones.

Cautionary words are offered here because a satisfactory shape in these areas is important to the overall appearance of the blade. Because of their shape and the coarseness of the early stones, it is easy to remove too much metal and permanently wear down, damage, or deform these areas.

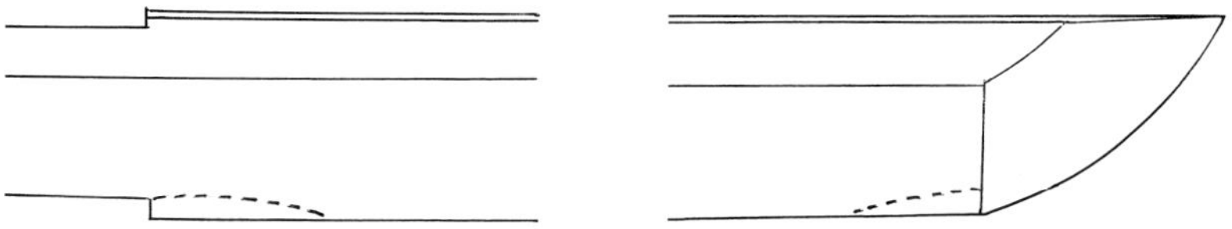

The hamachi (*left*), the area at which the cutting edge begins, and the edge near the yokote (*right*) are easily overpolished.

The area close to the hamachi (just above the tang) is being polished. Takaiwa applies more force when working this area because this part of the blade is thicker and often more metal must be removed. The edge is facing toward him and he is using the part of the stone nearest him.

Ji: Sides of the Blade (between Cutting Area and Ridge Line)

OBJECTIVE

To properly shape the sides of the blade, or, if necessary, repair and reshape them; to straighten the shinogi line.

STONE

Use a stone with a flat or rounded surface.

POLISHING METHOD

In area 3 (see figure on page 77), the polisher works his way along the sword a small section at a time in the same manner as he did when polishing areas 1 and 2. The polisher must produce a proper curve from the shinogi to the cutting edge. Polishing is done by moving the blade,

Takaiwa is shaping the sides of the ji and moving the sword along the lengthwise axis of the stone. The sides of the sword should have a slight curvature and not be too flat. Too much curvature, however, must be avoided. The stone markings on the sword will be about 60 degrees relative to the lengthwise axis of the sword. With each successive stone, the angle of the stone's marks will be closer to the lengthwise axis of the sword, and from the chu-nagura stage on, all the stones will be used in the direction of the lengthwise axis of the sword.

while keeping it flat, over the stones. The ji is polished in a series of small sections. Sometimes the curve of the ji is defined by as many as fifteen individual planes. With a new sword, it is possible to polish it at a diagonal angle to the stone from the early stages, in order to polish a wider area; for an old sword with a defined shape, the direction should be perpendicular to its lengthwise axis in order to have better control as one works on small sections. This will allow the polisher to repair any blemishes and improve the shape.

HINT

In the past, when good natural stones were hard to find, and synthetic stones were not as good as they are now, the ji was often worn down more than the ha, resulting in an overpolished or worn-away ji, which would expose the core steel of the blade. This came about because the steel of the ji is softer than that of the ha. Today however, due to better techniques and better synthetic stones, it is possible to polish the ha without overpolishing or wearing down the ji.

Kissaki: Point

OBJECTIVE

To produce a beautiful, natural curve along the cutting edge in the kissaki; to shape and form the sides and proportions of the point; and to position the ko-shinogi and the yokote line.

STONE

Use a stone with a flat surface.

POLISHING METHODS

When polishing a sword that already has an edge, the polisher places the kissaki near the right-hand edge of the stone he is using so that only the kissaki area is over the stone. The rest of the sword is held to the right of the stone. The kissaki is always polished in a direction perpendicular to the lengthwise axis of the sword, and the blade is moved in very small strokes against the stone.

A clean cotton rag is wrapped around a section in the center of the sword and provides a place for the polisher to firmly grip the sword with his right hand. The polisher's left hand can lightly grip the area of the blade just below the yokote and press the blade over the stone. The angle of the point against the stone is controlled by the right hand,

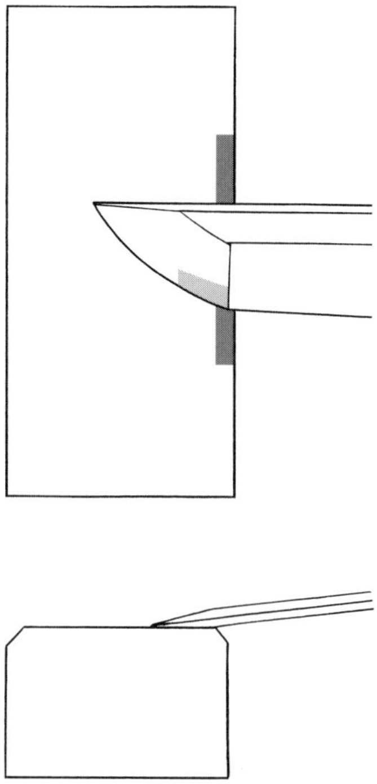

A

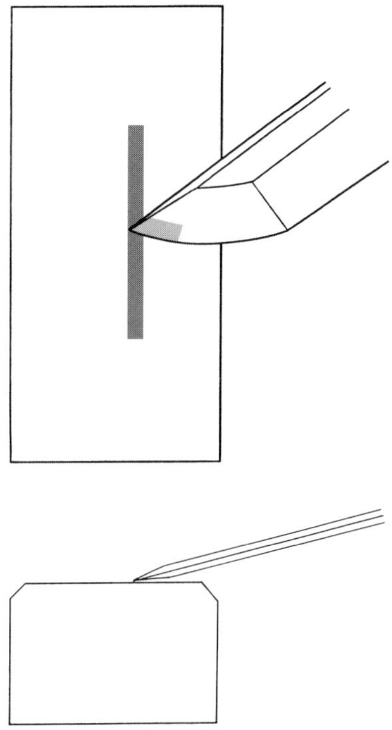

B

Diagram A shows how the area near the right edge of a stone is used when the edge of the kissaki near the yokote is polished. Diagram B shows how the right side closer to the center of a stone is used when the area near the tip of the point is polished. The blade is raised slightly so that the tip of the point can meet the stone at the correct angle.

Takaiwa is adjusting the shape of the kissaki. He is polishing the side of the point with the edge facing away from him. Notice the position of both hands on the blade. He is moving the blade along the stone in a direction perpendicular to the lengthwise axis. All polishing of the point (above the yokote line) is done in this manner.

which maintains a strong grip on the blade. The kissaki is polished with small up-and-down movements on the stone, back and forth away from and toward the polisher.

When polishing the kisssaki of a *shinogi-zukuri* sword (see page 173), the polisher begins at the edge of the kissaki, polishing the ha region (area 1 in the figure on the facing page). The width of the area being worked on during this process is very small and should be at the very most about ⅓ inch (7 to 8 mm) wide, since only a small part of the curved kissaki can be worked on at one time without flattening its shape. The polisher will work his way from the tip of the point toward (area 2) the yokote line along the mune and the ko-shinogi. He will then work the tip of the kissaki, before moving toward the yokote, just below the mune and up to the ko-shinogi line (area 3). The polisher will next work from this section along the ko-shinogi to the yokote (also area 3). Finally, he will polish the central section (area 4). If this area is flat or concave, the polisher may do little or no work on it, and the kissaki at this stage may look as if it were covered with a layer of fish scales.

After this, the entire point area will be shaped and formed properly up to the ko-shinogi line and the yokote, tapering from the mune to the cutting edge, and from the yokote to the point. This method is similar to that preferred by Takaiwa to polish the ha and ji. First, the area along the cutting edge is polished, then that closer to the shinogi line, until the shinogi line is reached. Finally, work is done on the central area. This type of slow, deliberate polishing of the kissaki, section by section, allows the polisher to form a proper and pleasing shape on each part, and to correct all of the irregularities in its shape, as well as to avoid polishing concave or depressed areas.

Takaiwa says this method of polishing the kissaki is extremely practical for most old swords because they do not have thick well-formed kissaki. This method can also be used to correct the shape of a rusted or damaged kissaki. There are many swords which have a kissaki with a concave surface that has depressions or dips. For damaged kissaki with a flat or concave surface in the center, it is important that the polisher not remove any metal from the center section, polishing around any depressed areas section by section. Once the shape is corrected, the kissaki is polished in a direction perpendicular to the lengthwise axis of the sword with the succeeding stones.

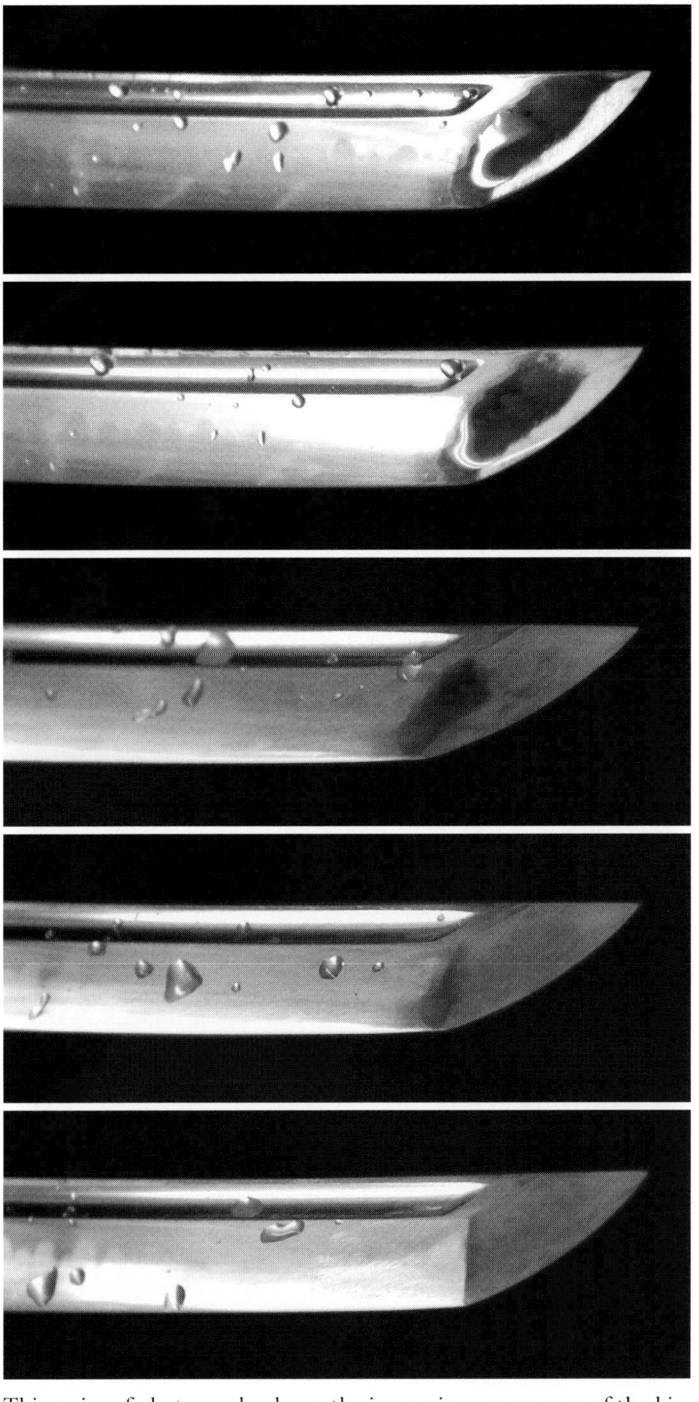

This series of photographs shows the improving appearance of the kissaki after each phase of polishing. It should be noted that the order of the areas being polished may vary depending on the shape and thickness of the kissaki.

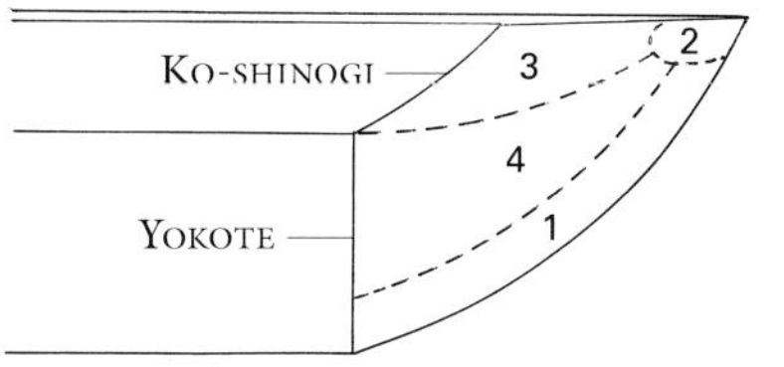

The numbers indicate the polishing order on the sections in the kissaki.

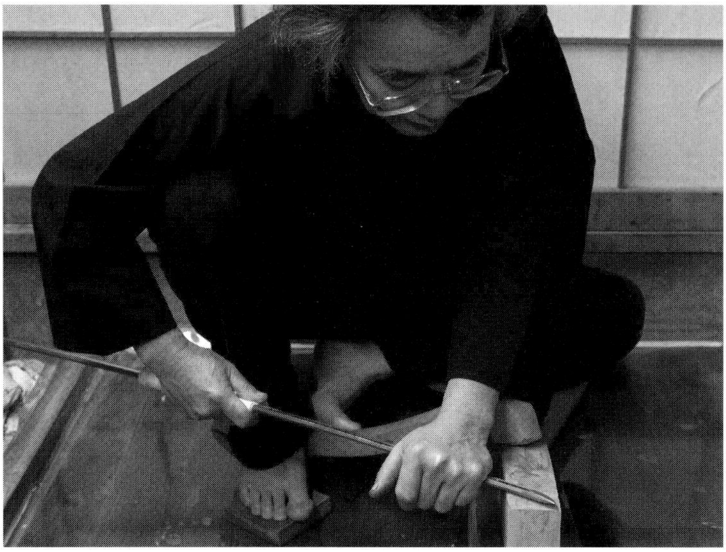

Polishing the kissaki near the yokote. The blade is moved up and down along the lengthwise axis of the stone. The point area between the yokote line and the tip of the blade is always polished in this direction.

Takaiwa is working on the top of the kissaki, moving the blade along the lengthwise axis of the stone. Notice that he has raised the tang end of the blade slightly higher than the stone to tilt the blade and make sure that the strongly beveled edge of the kissaki makes contact with the stone at an effective angle.

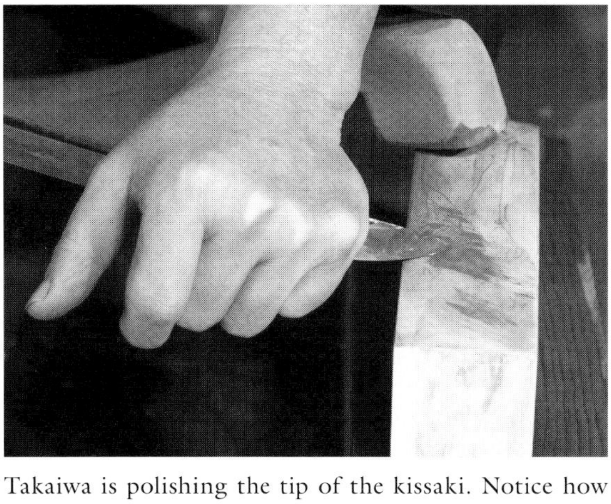

Takaiwa is polishing the tip of the kissaki. Notice how he holds the blade with one hand just below the yokote. When polishing the tip like this, the polisher should use the center of the stone rather than work on the edge. It is easy for the sword to slip off the stone if the edge is used.

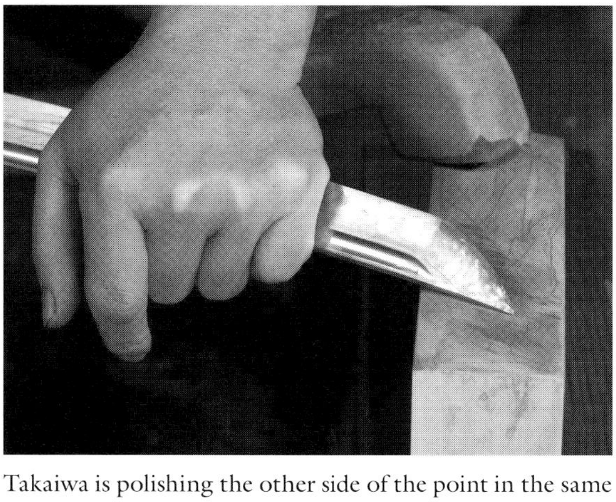

Takaiwa is polishing the other side of the point in the same manner, but with the edge facing him.

Hi: Grooves

OBJECTIVE

To polish inside the hi (see page 178 for illustlations).

STONES

A series of stones is used. These are, in the proper order, the arato, binsui, kaisei, synthetic nagura, and natural koma-nagura.

POLISHING METHOD

If the inner surface of the groove is uneven, rusty, or irregular, it must be polished and repaired. The polisher cuts a piece of synthetic polishing stone into a rectangular shape 3 by 4 inches (7 by 10 cm) and thick enough to fit into the groove. The polisher holds the stone in one hand and uses it to smooth out the inside of the groove to remove rust and irregularities and to even out the depth and surface of the inside of the groove. Small stones are made from each of the polishing stones, from arato to nagura. After the stones have been used, the next steps are performed with sandpaper followed by a *migaki-bo*, or burnishing needle.

A helpful aid to polishing with these small stones is to shape a small rectangular block of wood and cut grooves into it. The small stones are then fitted into the slots, and the wood block servea as a grip for the small stones.

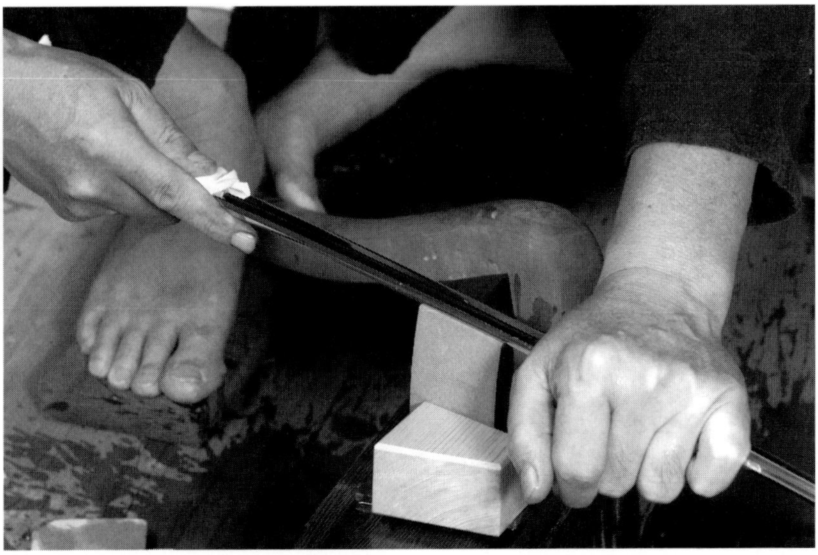

Polishers often use a slotted block of wood to polish the hi. A rectangular piece of synthetic polishing stone is cut to fit into the slots. The groove is polished with the stone by moving the sword lengthwise, with a very slight seesaw motion, back and forth, at widths of approximately 2 to 3 inches (5 to 8 cm) at a time.

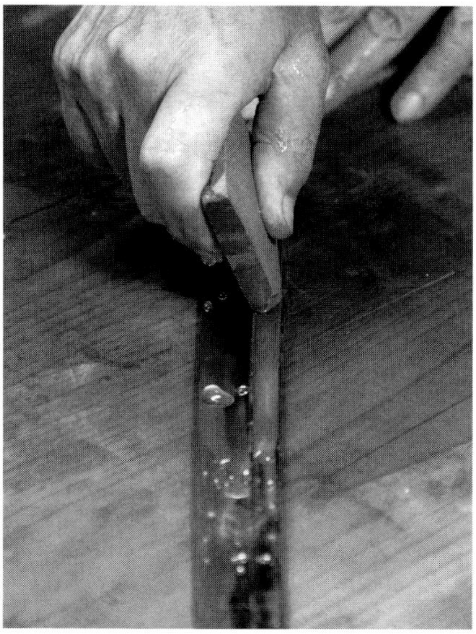

Takaiwa polishes the grove with a hand-held stone. He can distinguish the polished surfaces from unpolished ones by the color difference. The color of the latter is lighter.

SANDPAPER

A piece of fine sandpaper (No. 1200) about 2 by 3½ inches (5 by 8 cm) is cut and folded twice. The sandpaper is then wrapped around the end of a small wooden rod or dowel. A few drops of clove oil are added to the groove for lubrication. It is necessary to hold the sword firmly so it remains stationary. The sandpaper is dipped in the oil in the groove, then the bottom walls of the groove are sanded, working on about 2 to 3 inches (5 or 6 cm) at a time. The oil helps to make the scratches shallower and finer.

MIGAKI-BO

The migaki-bo is a hardened-steel burnishing needle used for the last stage of polishing the groove. For best results, the needle must be very smooth and have a fine finish. To prepare the migaki-bo for use, the polisher must make a thin groove in a small piece of wood. The groove should be about the same width as the point of the migaki-bo. The point is then smoothed and polished by rubbing it in the groove of the wood, lubricating the process with a paste of *tsunoko* (powdered horn and water). Once the migaki-bo is polished, it is used to polish the inside of the groove in the sword. It is placed into the blade's groove and rubbed over the entire inner surface of the groove. As the polisher works his way along the groove this way, the inner surface will become smooth, bright, and reflective. It is necessary to be careful not to over use the migaki-bo, since doing so will leave marks and dark patches inside the groove.

It is difficult to polish the inside of the groove with a stone and water because good contact with the concave surfaces cannot be made with a stone. However, there is no need to produce as fine a polish inside the groove as on the body of the blade, so in this respect the task of polishing the groove is easier than polishing the other surfaces of the sword.

It is a good idea to only polish the groove up to the kaisei stone. If the polisher continues with finer stones, rust will build up in the groove as the remainder of the sword is being worked on with the later stones and water.

The tools to burnish the groove with sandpaper (from left): clove oil, sandpaper cut and folded into an appropriate size, folded sandpaper, and a dowl.

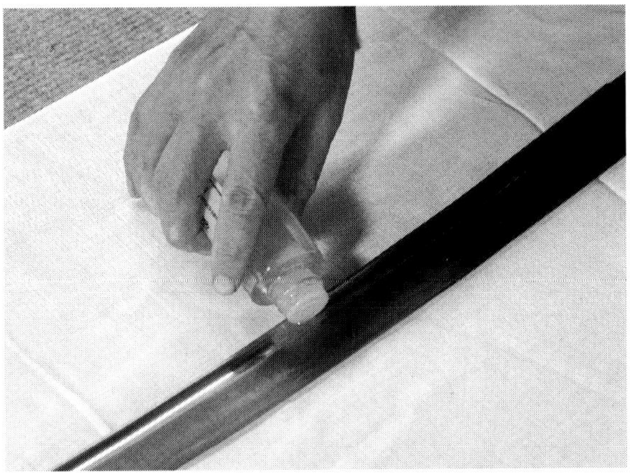

Takaiwa places drops of clove oil in the hi before he uses the sandpaper. Notice how the blade is placed on a clean soft surface to protect it while work is being done on it.

Takaiwa polishes the groove with sandpaper.

◄ Takaiwa uses a burnishing spatula to begin burnishing the groove.

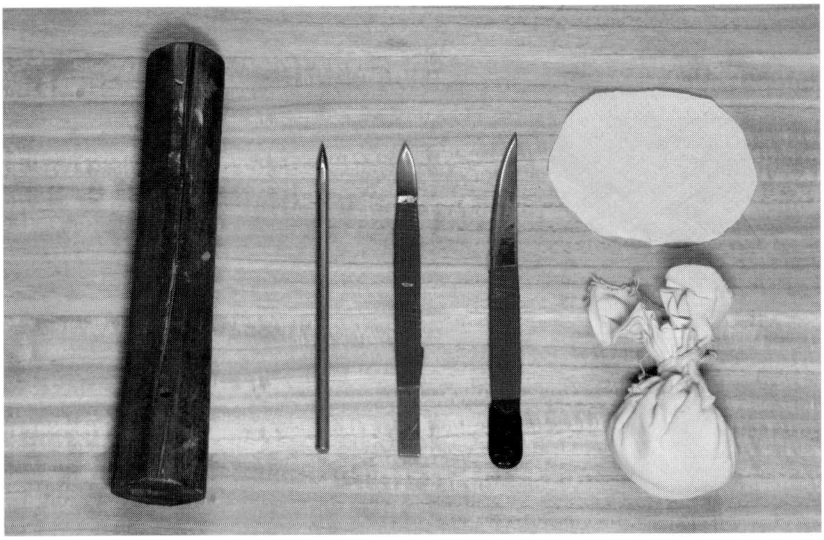

Tools used for burnishing work (from left): An old sword hilt with a groove in it used to polish the burnishing tools; a migaki-bo, or burnishing needle; two types of *migaki-bera*, or burnishing spatulas; a piece of cloth which is held between the polisher's fingers and the steel surface; and a cotton bag containing *ibota* lubricating powder.

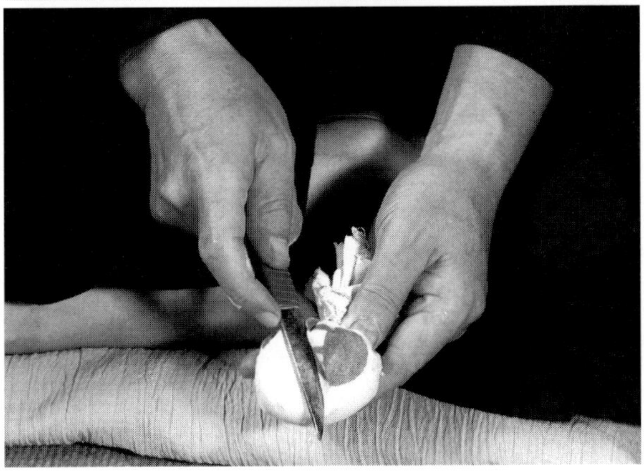

Before using the burnishing tools, Takaiwa rubs them back and forth in the groove of a piece of wood coated with tsunoko paste. He then rubs the tools with paste followed by ibota powder.

THE POLISHING STONES

The Koma-nagura Stone

OBJECTIVE

This stone is used to polish the entire length of the blade between the ha and the shinogi, shinogiji, and mune.

STONE

Natural koma-nagura stone, with a very slightly rounded, or convex, shape.

POLISHING METHOD

When polishing the kissaki with this stone, the method used is exactly the same as described for the rougher stones used earlier for the point area, which is polished in a direction approximately perpendicular to the lengthwise axis of the sword (this is true for all of the stones used to work on the point), or in a direction at a 90-degree angle to the direction at which the rest of the sword is polished. To polish the body of the sword, it is placed on the stone at an angle of 45 degrees, diagonally to the stone's lengthwise direction and is moved lengthwise over the stone using a very slight seesaw motion. Only about 3 to 4 inches (7 to 8 cm) of the blade is worked on at one time. The polisher applies force when pushing the sword forward (away from him) rather than when pulling it back. Takaiwa polishes the ha area in this way and uses an average of about 150 strokes across the stone for each 3-to-4-inch (7 to 8-cm) section. On the area from the nioi-guchi to the shinogi, he polishes each section of the same length using about 50 strokes. For the area around the nioi line, or around and on the boundary of the ha and the ji, respectively, Takaiwa uses about 100 strokes. This method helps produce a smooth continuous curve on the side of the sword.

CAUTION

When polishing the ji immediately under the shinogi line in a lengthwise direction, it is possible to polish a longer area with each stroke. However, this also means that the polisher has to be especially careful when polishing the ji just below the shinogi line: he must not polish above the shinogi line or disturb the contour of the line. One way to prevent damage to the shinogi is to keep the lengthwise strokes short.

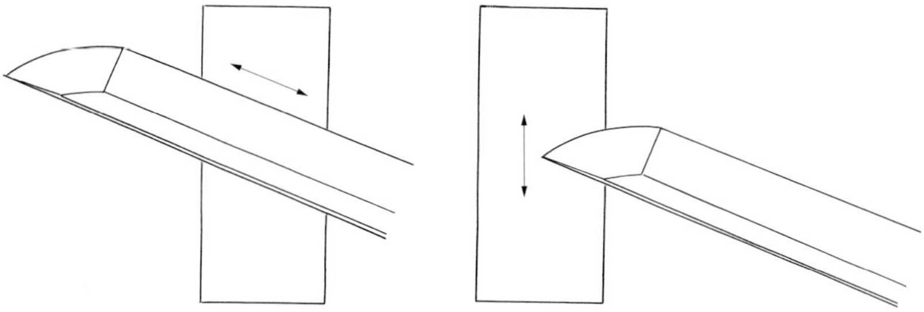

When using the koma-nagura and uchigumori stones, the sword is placed diagonally to the stone's lengthwise axis.

Takaiwa is polishing the ji using a koma-nagura stone. He holds the blade firmly near the tang, his right thumb pressing on the blade. The blade is placed diagonally against the stone and moved along the sword's axis with a very slight seesaw motion, back and forth. The wooden container in the foreground holds water with a small amount of soda in it to prevent rusting.

Takaiwa is polishing the ji. Notice how his right hand firmly grips the sword (over a rag) while his left palm presses down on the blade and his fingers curl under the edge to guide the sword.

The area of the ji below the yokote is being polished. The blade is moved back and forth in a lengthwise direction. Takaiwa uses the palm of his left hand to guide the blade along the stone.

The edge is being polished just below the yokote line. In order to even out and smooth the surface of the komanagura stone, Takaiwa rubbed it with a natural binsui stone before he began. The smaller binsui stone can be seen near Takaiwa's left hand. Rubbing the small stone over the large one smooths and cleans the latter's surface and produces the tojiru paste that lubricates the sword's movements over the stone.

The tang end of the blade is raised a little so that the tapered side of the kissaki along the edge can be polished.

Takaiwa corrects the surface of the koma-nagura stone with a rougher natural binsui stone after polishing the kissaki. The angled kissaki distorts the surface of the stone much more rapidly than do flat areas like the ji.

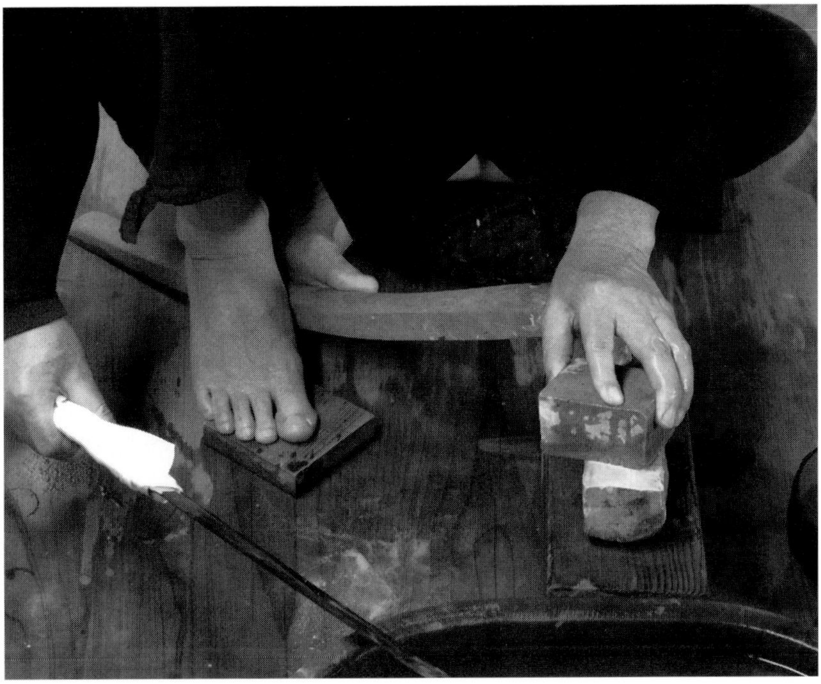

The Uchigumori Stone

OBJECTIVES

To remove the stone marks left by the koma-nagura stone and to bring out the details of the hamon and the texture and character of the jigane.

STONES

There are basically two stones used for the uchigumori stage: a soft uchigumori stone to polish the hamon, and a hard uchigumori stone to polish the ji. The surface of these stones should be flat. The edge of the sword is hard, so if a hard uchigumori is used to polish everything, it will simply slip on the hard hamon steel with little effect. In some cases, a hard uchigumori will produce a visible grain pattern, or *jihada*, in the hamon, which is undesirable. If a soft uchigumori is used on the soft ji, it will not produce a visible grain pattern, and the result will be a dull, featureless shine. In any case, the hardness of the blade's steel varies with each individual sword, so the polisher should have several different kinds of uchigumori in order to be able to select a stone that will work well with the sword at hand.

POLISHING METHOD

Polishing with the uchigumori stone involves two steps. The first involves polishing the hamon, and the second the ji. The polisher exerts force on the stone when pulling the sword back; the pushing motion over the stone should be simply to reposition the sword for the next pulling stroke.

Uchigumori ha-to: This is a soft stone mainly used for polishing the edge (ha) of the sword, and also for the ji. Most of the polisher's time is spent on the high areas between depressions in the hamon and on the edge of the sword. The sword is held with its lengthwise axis on a diagonal to the stone's length, and the sword is moved in a lengthwise

Takaiwa uses the uchigumori ha-to stone to polish the ha. He polishes one side entirely, from the hamachi to the yokote, before going on to the other side.

The beginning of the cutting edge around the hamachi is being polished, lubricated by tojiru paste.

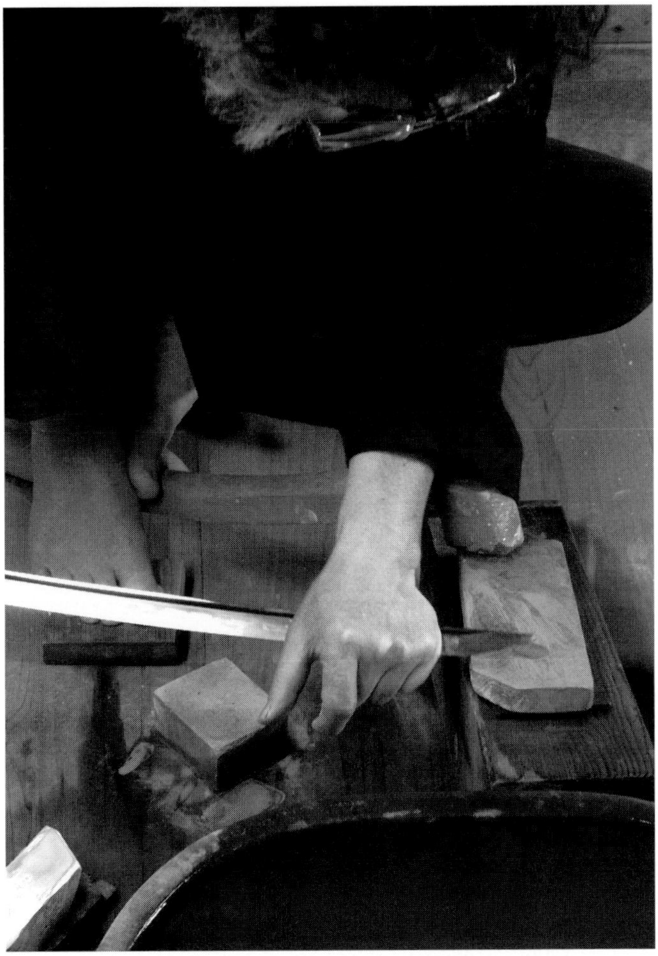

Takaiwa polishes the boshi. Again, the kissaki is being moved along the stone in a direction perpendicular to its lengthwise axis.

The area near the yokote line in the kissaki is being polished using softer uchigumori ha-to.

direction over the stone. The sword should not be moved with a seesaw motion as was done at the koma-nagura stage, but be kept flat. The place where the blade and stone come in contact should be thought of as a "point" rather than as a large "area." The area of the sword that can be worked on at any one time will vary depending on the curvature of the blade and the thickness of the sword. As a rule, an area of about 1½ inches (4 cm) is polished at one time at the very edge of the sword. As the edge is hard, the polisher uses a lubricating paste called *tojiru*; this is made by rubbing a small piece of uchigumori ha-to over the uchigumori polishing stone with some water between the stones. The rubbing action produces a thin bluish paste, which is applied to the polishing stone's surface. If the uchigumori stone is well chosen and used effectively, the hamon will be free of any scratches or marks and, if present, nie, nioi, sunagashi, ashi, or kinsuji details will emerge. The kissaki is polished with the uchigumori ha-to stone in a direction perpendicular to the sword's lengthwise axis, or at a 90-degree angle to the direction at which the rest of the sword is polished.

Uchigumori ji-to: This hard stone is used for polishing the ji and is used after the uchigumori ha-to. Once again, it is necessary for the polisher to select the appropriate ji-to stone. The uchigumori ji-to stone is also used to polish the shinogiji and the mune. However, in polishing the mune, the boshi area, which is very hard, must first be polished with an uchigumori ha-to stone. The goal of polishing with the ji-to stone is to bring out the jihada, ji-nie, utsuri, and chikei (if present) in the ji, and also to polish areas such as the shinogiji to produce a good balance in the color and appearance between the shinogiji and the ji. The shinogiji is usually either featureless and without grain, or rough-grained. Thus, polishing at this stage involves creating a balance in the appearance of two very different sword surfaces, the shinogiji and the ji.

There are three points to keep in mind when polishing with this stone:

1. Before starting to work with the uchigumori stone, all the water used up to this point should be discarded. The water bucket and the work area should be cleaned thoroughly. No dust or particles from the coarser stones should remain in the working space as they can cause scratches on the blade during these last polishing stages.

2. Use only natural stones to produce the tojiru paste. A tojiru produced with a synthetic stone is not sufficiently fine and will lead to visible scratches on the surface of the blade.

3. The mune and hi can be polished with sandpaper and a burnishing needle after using the synthetic nagura stone, so it is not necessary to use uchigumori stones to polish three parts. Because the hi are narrow and concave, it is not necessary to be concerned about the texture and appearance of their interior surfaces.

A pair of old koma-nagura taken from a quarry in Kyoto over a century ago.

Chapter 7 FINISH POLISHING

Jizuya

Takaiwa does not use a hazuya stone to polish the ji because he finds that it has little or no effect on the blade. However, many polishers do use the hazuya stone on the ji after using the uchigumori ji-to stone. The hazuya, prepared and used in the same manner as the jizuya stone, is used to go over the entire blade after using the last uchigumori stone. Takaiwa chooses not to use the hazuya so details about this step are not illustrated. It is worth noting that when hazuya are used, it is followed by jizuya. Since this polishing description is following the work of Takaiwa, the hazuya stage is not shown here, but the preparation and use of the hazuya follows the method described here for the jizuya stones.

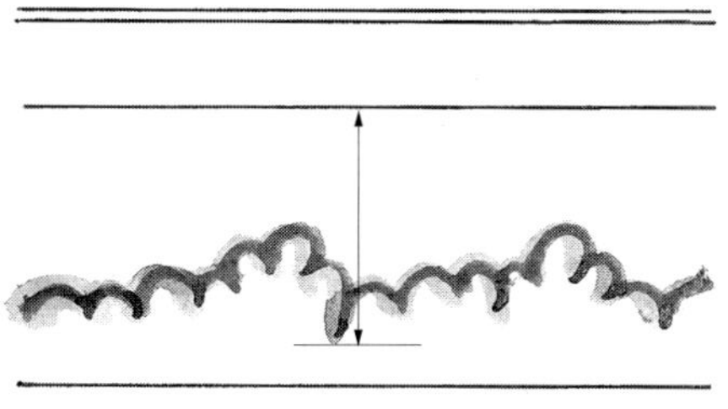

The area of the ji to be polished with the jizuya is between the shinogi line and the lowest part of the hamon.

The two stones on the left are *uchigumorido*, used for making hazuya; the ones on the right are narutakido, used for making jizuya.

The order in which the stones are used is as follows: uchigumori jito → jizuya → nugui → hadori. At the jizuya stage, it is advisable to add a small amount of washing soda (sodium carbonate) to the water to inhibit rusting. Because these steps are time consuming, the surfaces of the sword not being worked on directly are exposed to water and easily rust if preventative measures are not taken.

OBJECTIVES

To remove the marks on the sword left by the last uchigumori stone, and to bring out the texture, characteristics, and fine details of the blade surface.

STONE

Narutakido. One should use the best-quality narutakido available. The hardness of the narutakido stones varies, and the polisher must match the stone's qualities to the hardness of the sword's surface. When used for the jizuya stage of polishing, which is the beginning of the final polishing process, a good narutakido stone will produce a dark-colored paste, or tojiru, when it is rubbed on the sword with water, while a bad stone will produce a thick, muddy-looking paste.

As a general rule, when the sword's surface is hard, the polisher should use a softer jizuya; when the sword's surface is soft, the polisher should use a harder one. In either case, the polisher should have a stock of at least ten different types of jizuya on hand from which to select.

MAKING THE JIZUYA

1. Starting with a wafer-thin piece of narutakido (which has been split from a larger piece of stone with a chisel), it is ground down on rough stones (called omurado and aoto) until it is about 1⁄12 to 1⁄8 inch (2 to 3 mm) thick.

2. The ground-down narutakido is covered with a mixture of traditional Japanese lacquer and a synthetic resin. A piece of coarse Yoshino paper is attached to one side of the stone using the lacquer mixture, and left to dry for a day.

3. When the Yoshino paper and lacquer are dry and have adhered to the stone, the stone is ground down again on a coarse omurado, using water, until it is 1⁄60 to 1⁄50 of an inch (0.4 to 0.5 mm) thick in the case of softer narutakido, or 1⁄80 inch (0.3 mm) thick if it is harder.

4. The paper-backed narutakido is then cut into small squares with a knife or razor blade. A softer narutakido is cut into squares about ½ x ½ inch (12 by 12 mm), the medium type into squares ⅜ by ⅜ inch (10 by 10 mm), and a harder one into ⅓ by ⅓ inch (7 x 7 mm) squares.

Takaiwa uses a thin chisel or wedge and hammer to cut thin sections from the narutaki stone.

A thin section of narutakido is ground down on an omurado stone with water until it is uniformly about ⅛ inch (3 mm) thick.

5. Next, using a fine knife, a grid of shallow lines is cut into the top of the stone.

6. The grid-lined jizuya is then placed on the fingers of one hand with the paper side facing upward, and a wooden dowel or rod of about ⅝ inch (1.5 cm) in diameter is rolled over the jizuya once. This opens up the grid of fine lines scored into the stone and makes the jizuya malleable and flexible.

The narutakido is next ground on an aoto stone, which is much finer than the omurado. It will be worked on until it is about 1⁄12 inch (2 mm) thick. One of the reasons for using the aoto stone is to ensure a very smooth surface.

The four uchigumorido pieces on the left that are to be used as the hazuya have been cut with a power grinding wheel, while the pieces of thin narutakido on the right are to be used to make jizuya.

A lacquer-and-resin mixture is next spread evenly on the surface of the jizuya. This modern mixture from a tube dries in one day. Traditional lacquer would take about a week to dry.

A piece of Yoshino paper is carefully placed over the jizuya.

Takaiwa cuts a piece of jizuya into strips with a knife.

Takaiwa cuts a strip into lengths suitable for use.

Three different pieces of jizuya resting on the blade. The largest one is the softest and the smallest is the hardest. The largest is used to remove the stone marks from the uchigumori. The blade itself has been polished with the uchigumori ji-to stone, and all the details and outline of the hamon are clearly visible. In spite of having used the ushigumori, the surface and texture of the sword still appear somewhat rough and harsh, and the first soft jizuya will remove these last signs that the large stones were used. The smaller jizuya stones are used in succession to bring out and enhance the jihada.

A piece of jizuya with a series of grid lines scored into the stone. It will be broken along the gridlines before being used.

Takaiwa places the scored jizuya on the fingers of one hand with the paper side facing upward, then gently rolls a dowel over the jizuya to create a grid of cracks in the stone. The cracks will make the jizuya even more flexible and able to follow the contours of the sword.

USING JIZUYA

1. The blade must be kept very clean. If dust comes between the blade and the polishing stone, or the blade slides on a surface, new scratches will be created. This means that when repositioning the blade during these final polishing steps, it must be placed on clean working surfaces. Any place the blade is to rest during this stage must also be covered with a clean cotton cloth.

2. To begin working with the jizuya, prepare by placing a clean piece of cloth on one knee, and place the sword on it (taking care not to slide the blade over the cloth or over any other surface). The blade should rest on one knee with the cutting edge of the blade facing away from the polisher. The blade must be stable during this process. When the polisher switches from working on one side to the other, he rotates the sword 180 degrees end over end, so that the cutting edge still faces outward. This requires the polisher to

Takaiwa is polishing the ji with a very thin jizuya. He adds a few drops of water from time to time if the section he is working on is too dry. Because the ji is soft compared to the ha, it is good to use the tojiru liquid produced while polishing with the jizuya.

use his other hand to continue polishing. Assuming the polisher started by gripping the hilt with his left hand, after flipping the blade end over end he must grip the hilt with his right hand, and polish with his left. He moves along the face of the sword lengthwise, working on a 2-inch (5-cm) section by moving the jizuya back and forth along the length of the blade. The jizuya is then moved along the blade to the next section and used in a similar manner, working from the edge up to the shinogi, and then back down to the edge. The polisher must move the sword as he moves from one section of the blade to the next with the jizuya; the area being polished must be positioned directly below the polisher's eyes so he can polish in a comfortable position. Once again, it should be emphasized that the sword should not be moved over the cloth to a new position, but should be lifted up, moved to the new position, and then set down again directly on the cloth to avoid scratching the blade. To work on the other side of the sword, the polisher will grip the hilt in his right hand, position the edge to face away from him, and use the thumb of his left hand to work with the jizuya.

In the earlier stages of the polishing process, when the large stationary stones are used, the polisher can hold the sword with the edge facing toward or away from him, whichever is the most comfortable. However, at the jizuya stage, it is imperative that the edge of the blade faces away from the polisher, so that he can gauge the effect of the jizuya stones, position them properly, and carefully observe the progress at each step of the polishing process.

With the jizuya, the polisher starts polishing with the larger and softer jizuya stones. When using the first jizuya stone, the polisher adds a few drops of the tojiru paste to the portion of the blade's surface as a lubricant. At this stage, the jizuya is used to remove the stone marks left from the uchigumori stones. The polisher goes over the entire surface of the blade with the first jizuya stone, and then works with the smaller and harder jizuya stones in sequence, using tojiru and water as lubricants. If used properly, the stones will bring out the grain pattern on the blade's surface. With the hardest jizuya stones, it is a good idea to use a smaller concentration of soda in the water so that the jizuya stones are not too slippery and hard to control.

For the final stages of the polishing process it is a good idea to have a separate small (wooden) bowl or just a cup of water with sodium carbonate to prevent rusting. The concentration of the soda should be about 1 ounce (28 grams) to about ¾ cup (180 ml) of water although the optimum concentration of soda varies depending on the stage of the polishing.

The shinogiji will be further polished later, using burnishing spatulas, so it is not necessary to spend as much time or effort on the shinogiji as on the ji.

When using the softer jizuya stones, it is a good idea to prepare some extra stones, so any that wear out during the course of polishing can be replaced immediately. Polishing with worn out jizuya requires more force and may lead to an uneven finish.

Some suggestions to help use the jizuya stones effectively are:

1. Use a soft jizuya for swords with a coarse surface. If you use a hard jizuya in this situation, the defects or irregularities on the blade's surface will simply become more pronounced. The use of

a soft jizuya on a rough or irregular surface can help reduce or soften the visual impact of any defects or irregularities.

2. When polishing, the polisher should keep the area being worked on directly beneath his eyes, and should constantly adjust the position of the sword on his knees to maintain a comfortable working position. The sword must also be kept steady during this process.

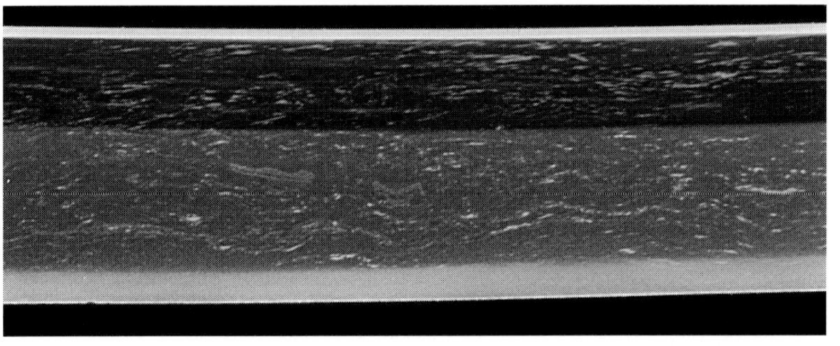

A

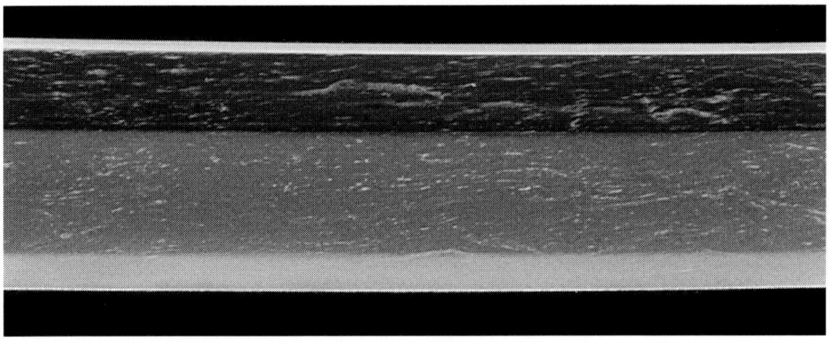

B

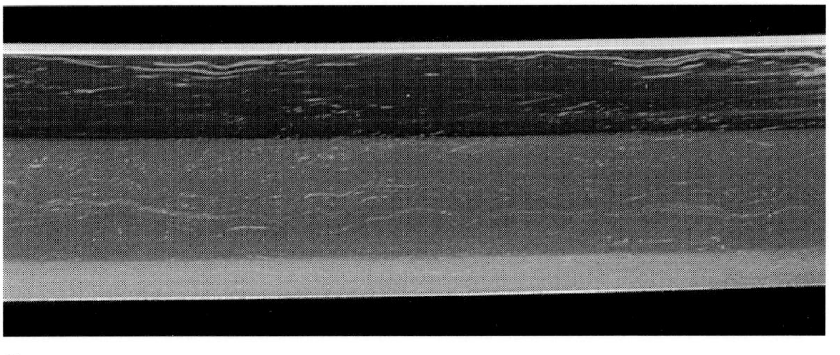

C

A blade with a rough surface is polished with three different jizuya stones. Photo A shows the surface polished with a hard jizuya stone; as a result, the surface has become rough. Photo B shows the results of polishing with a stone of medium hardness, and photo C shows the effect achieved with a soft stone. The rough surface looks smooth.

3. In polishing the ko-shinogi area, it is necessary to remove any marks produced by the nagura and uchigumori stones previously used in this area. Scratches may be extensive because the ko-shinogi is a difficult area to work and the force exerted is often strong.

4. In the area of the ji immediately in front of the yokote, the polisher should use a thin jizuya. Because the yokote line physically sets off and defines the point area, and is slightly higher than the surface of the ji, the polisher could easily polish the yokote line down to the surface of the ji if he uses a thick piece of jizuya. Using a thin, flexible jizuya stone around the yokote will help to protect the yokote.

5. The part of the ji directly below the shinogi line is very difficult to polish. Often it is left almost unpolished, which produces a whitish appearance in the area below or on the shinogi. This happens because polishers are afraid of eroding the shinogi. However, if a polisher wants to do a proper job, he must make the effort and polish this area. The polisher should avoid trying to follow the curve of the shinogi line exactly. Instead, he must bring the edge of the jizuya up against the shinogi line and polish in a straight line *at a tangent* to the contour or curve of the shinogi.

Kanahada Nugui

After all of the work with the various polishing stones is finished, one of two nugui preparations is used: either kanahada or sashikomi. For kanahada, a mixture of finely ground iron oxide and clove oil is made. Rubbing the blade with this mixture softens and improves the appearance of the blade's surface. Each nugui mix produces a different finish. The decision about which to use depends on the polisher, the sword, and the area to be polished.

OBJECTIVES

Even after using all of the stones to shape and bring out the details in the sword's ji, the appearance of the blade will be a bit bright and harsh. The use of nugui is intended to even out the color and to soften the appearance. Kanahada nugui is applied to the ji, shinogiji, and mune. The other important function of nugui is to darken the ji. A darker ji will provide a good contrast for the lighter hamon, and allow the polisher to make the hamon much more visible and clear.

IMPLEMENTS

Kanahada mix, metal flakes, Yoshino paper, and cotton wool are required.

When steel is heated to a high temperature, black flakes form on the surface of the metal as it cools down, and fall off when the metal is hammered. These flakes are the basic material for making kanahada mix. The flakes are placed on a ceramic plate and heated over a flame for more than twenty-four hours. The flakes are then ground up and heated in a pot. Next, they are ground in a mortar and pestle to a very fine powder. There should thus be no large particles left in the iron powder, since they would scratch the sword. Several drops of clove oil are added to the powder to make the kanahada mixture.

METHOD

Before use, the mix must be filtered through several layers of Yoshino paper. Yoshino paper is stacked or folded until there are eight to ten layers. The corners of the paper are twisted together to form a shallow dish. The polisher dabs some of the mix on the paper with his finger. Once nugui has soaked through the paper to the bottom, the bottom of the paper stack is used to apply the mix to the sword. The paper filters out the larger particles of iron oxide.

Two rectangles of cotton wool about 1¼ by 5 inches (about 3 by 12 cm) are used to rub the mix over the blade. After use, the cotton can be stored in an airtight container to prevent drying, and be used in the next step to remove or adjust the hadori finish.

The sword is held in the same manner it was held for the jizuya stage, with the edge facing away from the polisher. The polisher begins with the cotton under his thumb and rubs the blade with the mix. A small section of about 3 inches (7 cm) is worked on at any one time. The cotton patch is then moved up to the next section and the process is repeated. The polisher starts at the base of the blade just above the tang and works his way along the length of the blade up to the yokote. The nugui is worked from the edge to the shinogi and back. It is them worked over the next section of the blade. The mix is applied to the entire length of the sword, working up to the yokote and then back down to the machi again. (The kissaki or point will be finished later using a tool called a *narume*.) The polisher then repeats this process with a second, clean piece of cotton cloth. The hamon, which should have become clearly visible after the jizuya stage, will usually become darker once the mix is applied. In fact, the color of the hamon will probably be the same as that of the ji. The polisher can adjust the color

From the left: folded cotton wool pads; kanahada mix on Yoshino paper; and kanahada mix in its container. Note that the corners of the Yoshino paper stack are twisted in order to hold all of the layers together.

Kanahada flakes. A piece of iron or steel is heated to a red or yellow color and allowed to cool. As it does so, black scales form on the surface of the metal. If the metal is hammered, the scales flake off. These flakes are collected and ground into the fine powder combined with clove oil to make the kanahada mix.

Ground kanahada (left) and ground *jitekko* (magnetite) for sashikomi nugui.

of the sword by altering the amount of mix he applies to each section of the blade. More mix and more rubbing will darken the blade. Generally, it is best to put more mix on the blade near the hamon boundary, where the steel is very hard, and less toward the shinogi, where the steel is softer. It is difficult, but very important, to distribute the mix well to obtain the best results. That means it is necessary to be able to judge the quality and character of the sword's jigane, or steel surface. If the jigane is weak and soft (with a low carbon content) or coarse (frequently polished with the core steel showing through in places), the blade can quite often end up being too dark or display an irregular appearance after the mix has been applied. In such cases, the polisher can use a denser cotton wool patch when rubbing the mix over the blade's surface; the denser cotton will reduce the pressure on the sword's surface and lighten or soften the effect of the mix. The ideal appearance of the blade after the kanahada mix has been applied is a dark smooth surface. If too much force is applied, the color of the steel will go from a dark gray to a brownish appearance, which is not desirable.

Some other points to remember when working with nugui are:

1. With a regular jigane, but with a complex hamon (for example, one composed with a *juka-choji* and *sanbon-sugi* pattern), when it is difficult to produce a good finish using the *hadori* method (explained later), or when the hamon is beautiful enough as is, the polisher sometimes does not use kanahada mix, but will use an alternative finishing method called sashikomi nugui (see next section).

2. The appearance or color of the sword should be a dark gray, but not too dark or black. If the steel surface is too dark, then after the hadori finish the hamon will appear too white and bright against the dark ji. Thus, at this stage, the polisher has to consider what the final appearance of the polished sword should be.

Takaiwa uses his finger to daub the kanahada mix onto the Yoshino paper pad.

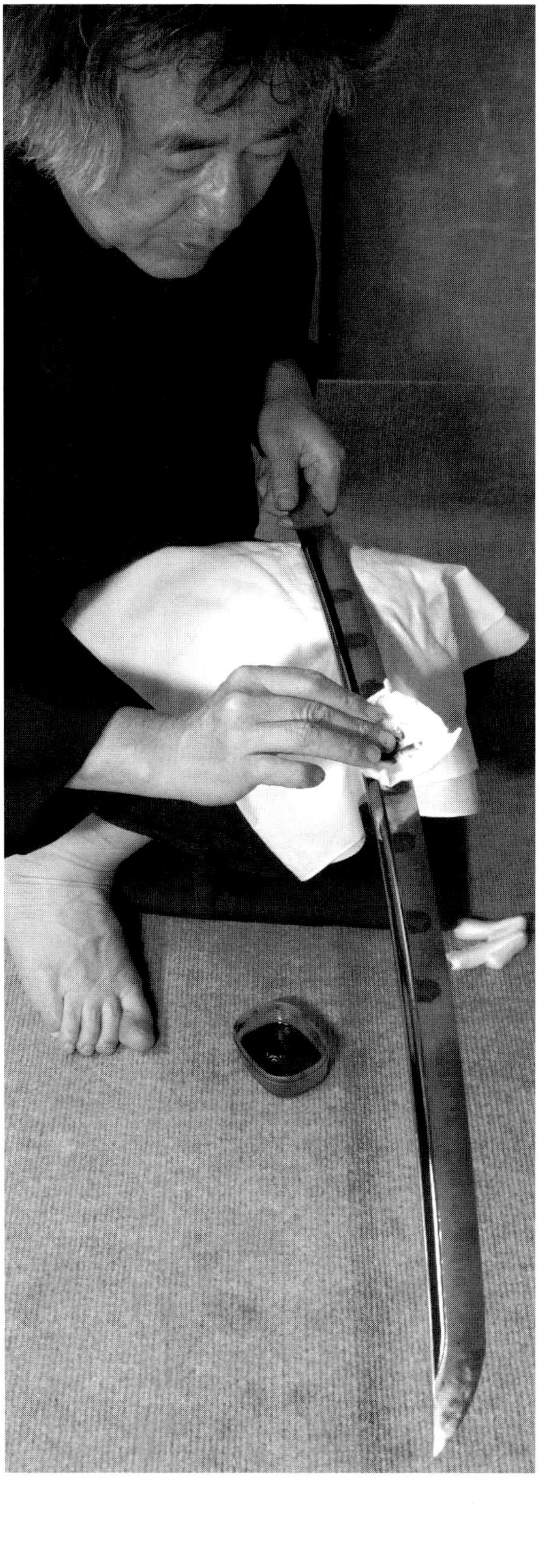

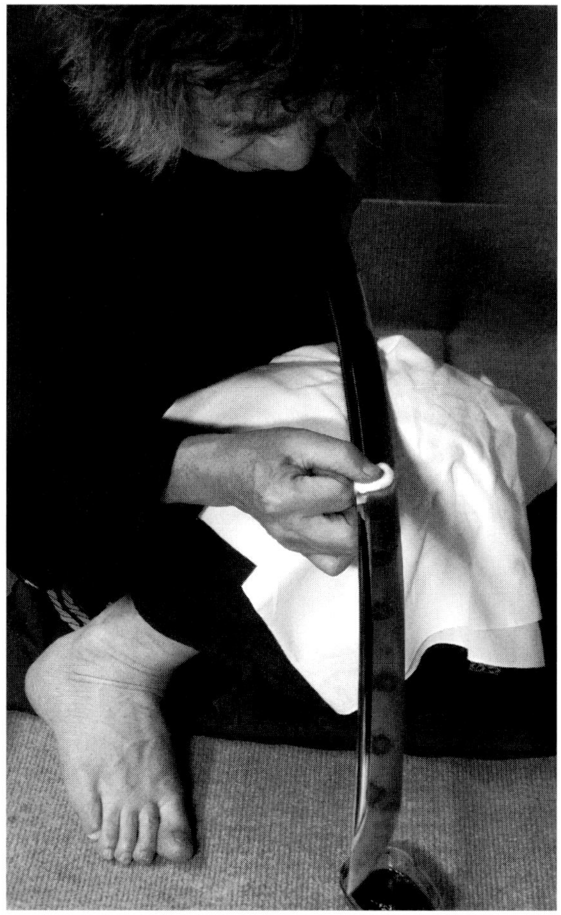

Takaiwa applies kanahada mix to the blade's surface. Notice that this is now very evenly finished and smooth, and that the ji and hamon are very clear after the jizuya stones have been used. The kissaki still appears rough and will be finished later in the narume stage.

Takaiwa uses folded paper to filter and apply the kanahada mix to the middle of the blade.

Sashikomi Nugui

OBJECTIVES

To finish polishing the sword in a manner that will produce a distinctive hamon, without working on the *nioi-guchi* (hamon boundary), the polisher can use sashikomi mix rather than the more commonly used kanahada mix. Sashikomi mix is the preferred finishing method when (1) the sword has a particularly complex and intricate hamon (for example *choji midare* or sambon-sugi; see page 183); (2) the hamon is already quite clear and well formed and there is no need to enhance it further; or (3) the hamon is so broad that the top of the hamon approaches the shinogi. In a case like this, using an ordinary hadori finish on the sword can result in most of the ji being white and overwhelming.

A sashikomi nugui finish will selectively lighten the hamon, but only inside the nioi line or hamon boundary, and effectively reduce the area of the white hamon to produce a more pleasing appearance on the sword. This type of finish will also lighten or emphasize only the area inside of the hamon and not affect the ji.

MATERIALS AND TOOLS

Magnetite (jitekko), clove oil, and cotton wool. Add finely ground magnetite powder to clove oil to form the sashikomi mix. It is not necessary to filter the mix through Yoshino paper. The cotton wool is the same as was used for kanahada nugui.

Finely ground magnetite powder is mixed with clove oil to make the sashikomi mix.

METHOD

The order in which the blade is finished is as follows: jizuya → hadori → sashikomi nugui. The hadori step used with sashikomi nugui is different from the hadori step (described in the following section) used after kanahada nugui. With sashikomi nugui, the polisher is not adjusting or modifying the appearance of the hamon in any way, but simply polishing the existing hamon to bring it out as faithfully as possible. There are two types of hadori polishing used in conjuction with sashikomi nugui. One polishes the inside curve of each wave. The other consists of quick polishing from the top of the hamon to the bottom. After hadori, boiling water is poured over the blade to heat it.

Then the sashikomi nugui is used. The polisher dips his fingers into the sashikomi mix and dabs it directly onto the surface of the ji over 3-inch (7-cm) sections then rubs it over the ji with the cotton wool. The sashikomi mix is applied over the entire blade (including the hamon) except for the kissaki. The ji will darken, but the ha will remain the same. (In using kanahada-mix, the hamon does not have to be considered, only the ji. This is because the final polishing steps to enhance the hamon will be done at the hadori stage after using the kanahada nugui.)

Several points should be noted:

1. Sashikomi nugui should not be used for swords with a rough jigane (when the shingane, or core steel, is exposed in the ji), or with a surface that has nicks, gouges, or depressions (or where the jihada shows separations between layers). If sashikomi nugui is used, it will highlight the defects.

2. If the polisher plans to use sashikomi nugui, he should avoid using a hard uchigumori ji-to stone during the polishing in favor of a stone of medium hardness. The sashikomi mix will enhance the jihada. Using a hard uchigumori ji-to will result in an exaggerated jihada.

In order to use the sashikomi mix effectively, the blade must be heated, so boiling water is poured over the blade.

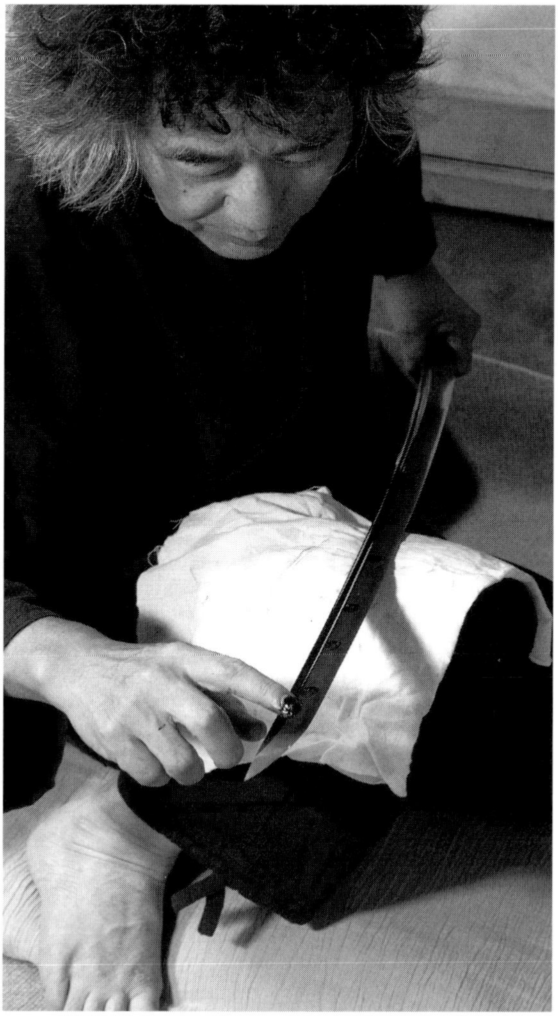

Takaiwa rubs the blade with cotton wool and sashikomi mix. Notice that the outline of the hamon is very clear and a lighter color, which now contrasts with the ji. The nioi line defining the hamon area is quite clear and there is a distinct contrast between the ji and hamon. Hadori work will not be necessary to further enhance the hamon

Takaiwa uses his finger to spot the blade with sashikomi mix.

Hadori

OBJECTIVE

To polish the hamon. When used with kanahada nugui, the idea is to bring out and lighten the hamon against the dark background of the blade's ji, and to produce a contrast between the hamon and other parts of the blade's surface.

TOOLS

Hazuya. A hazuya stone is made by chipping off slices from an uchigumori stone with a hammer and chisel. These slices are ground on a flat aoto stone (using water) until they are about 1/80 of an inch (0.3 mm) thick or less.

Takaiwa cuts hazuya stones into appropriate sizes for use. The stones are made in the same manner as the jizuya stones, but from uchigumori stones rather than narutakido stones. Here a knife is used to cut them into small squares.

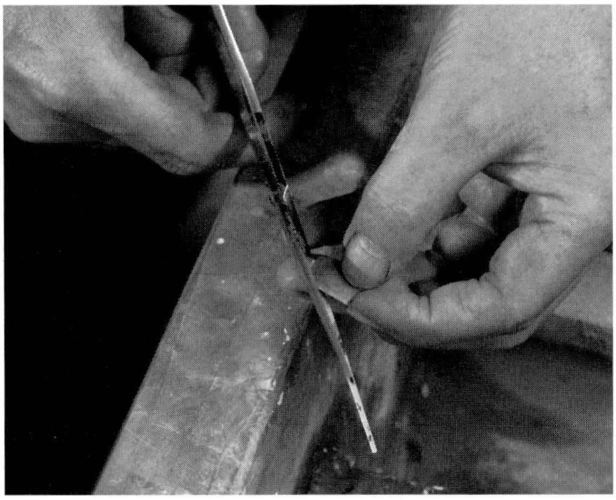

The hazuya stones are cut into ovals, or (as shown here) squares with rounded corners.

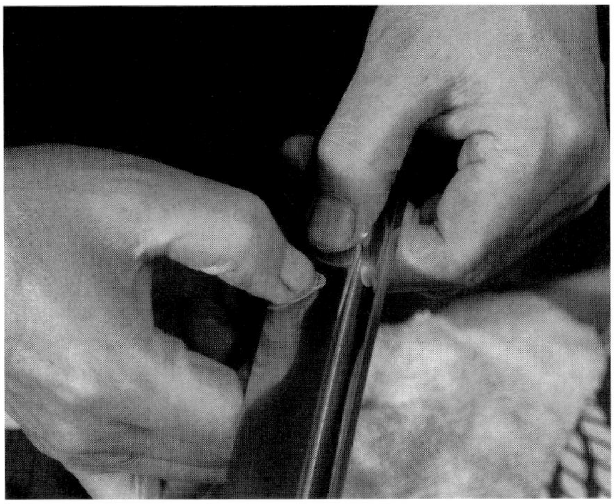

Takaiwa polishes a small section of the hamon using a hazuya stone and tojiru paste.

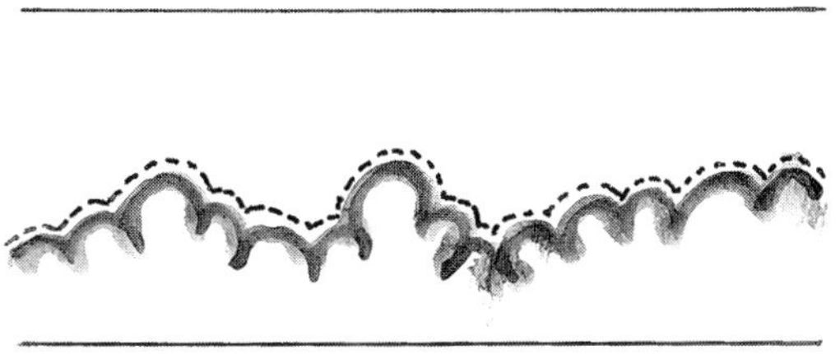

The dotted line shows the area in which the polisher uses the hazuya when he emphasizes the hamon boundary. The area slightly over the hamon boundary into the ji will become very light and produce a strong boundary on the ji.

► Here, Takaiwa is using the hazuya to begin polishing and whitening the hamon. The white area on the blade is a single loop or wave of the hamon, and it is covered with tojiru paste. The tojiru lubricant is necessary because the ha is a hard area and the stone does not work well without tojiru liquid. Notice that the sword is resting on clean cotton cloths and that Takaiwa is polishing the hamon from the edge side, with the edge facing him.

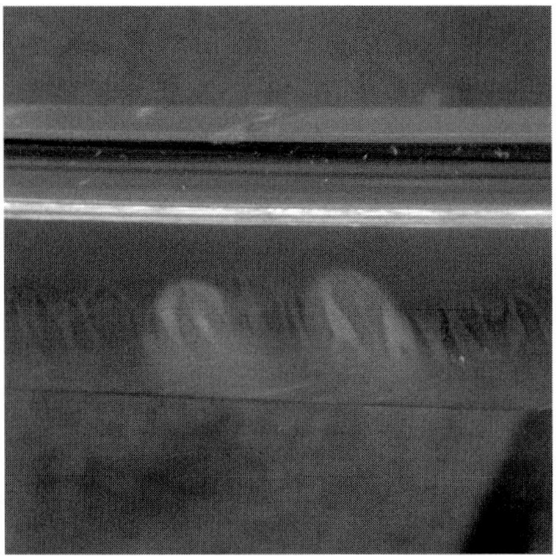

Takaiwa has used the hazuya on two waves of the hamon but has skipped one wave between them.

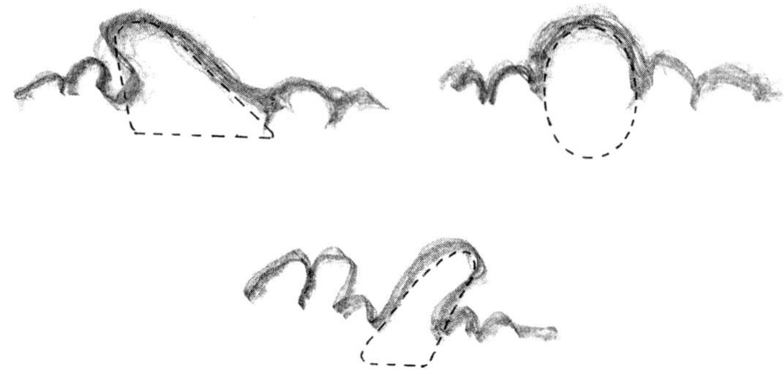

The size and the shape of the hazuya that Takaiwa uses are almost the same as the "waves" of the hamon. With small-sized hazuya, the polisher produces a quality finish by moving the small stone fragments within each wave in minute strokes.

The slices are painted with a lacquer-and-resin mixture. A piece of Yoshino paper is placed over the wet lacquer to provide a backing. The lacquer should saturate the paper. The grain of the paper must rise above the lacquer to provide a rough surface for the polisher's fingers to adhere to so that he can guide the stone when it is being used. The stone-and-paper combination is left to dry for a day. Before use, the hazuya stone is cut to a size that matches the average width of the hamon to be worked on. Sword polishers usually cut these stones into an oval or rectangular shape. If the hardness of the hazuya does not complement the hardness of the blade, the polisher will not be able to polish or whiten the hamon to the desired degree. In general, the darker the hazuya, the harder it is.

METHODS

Different sword polishers have their own methods for working with the hazuya stones. The Fujishiro school customarily holds the sword with its cutting edge facing the polisher, working from the cutting edge. Most other polishers work with the sword's mune facing them and polish from the mune surface toward the hamon.

Takaiwa, who trained in the Fujishiro tradition, holds the sword's edge toward him. The first step is to rub a piece of hazuya on an uchigumori stone with water (containing sodium carbonate) to produce a tojiru paste. The tojiru is spread on the portion of the hamon which will be worked on in order to lubricate the movement of the hazuya over the blade's surface. As a rule, the stone is used by moving it only forward and backward over the blade's hamon along the blade's lengthwise axis, and never diagonally across the blade. The stone is used from the very edge of the blade up to the nioi line defining the hamon (the nioi-guchi). The polisher should use the stone by working up to the top of the hamon and following the basic line or contour of the hamon. However, the polisher can emphasize the nioi-guchi by going up and slightly over this line into the ji, an action that can produce a strong white line on the surface of the ji, which will emphasize the appearance of the hamon's shape.

Placing the hazuya stone under the ball of his thumb, the polisher moves the stone back and forth over the hamon section on which he is working. The polisher moves the stone along the sword's lengthwise axis, working his way from the cutting edge up toward the ridge, a section at a time. In order to cover a complex shape, the stone is worked up to and slightly over the nioi-guchi, and centered under the highest point of a *choji* or *gunome* wave (see page 183). The individual waves of the hamon must be worked on one by one, and only by polishing in a lengthwise direction. The polisher can work along an imaginary line going from the edge of the blade to the highest point of the wave. He will use small back and forth movements of the stone near the narrow top of the wave. As he works his way down toward the edge of the blade where the wave becomes wider, he can use longer and longer back and forth strokes of the stone to cover the entire width of the wave. Working this way, the polisher can slowly polish and whiten one wave at a time, and then continue on to the next one. The hadori stone should never be moved in a direction perpendicular or diagonal to the blade's lengthwise axis. Takaiwa recommends polishing the hamon by working on alternate waves, then going back and working on the omitted waves. He prefers this method because he finds that he can acquire a better sense of the contours of the individual hamon and produce a well-balanced, evenly polished hamon.

For a *hira-zukuri* blade, the hadori used in the kissaki area is the same as the hadori used in the shinogi-zukuri blade. The narumedai (see page 124) is not used until the tip of the kissaki is polished. The outline of the hamon should be even and whitish. The shape and size of the hazuya is determined by the size and form of the boshi or the hamon in the kissaki. Takaiwa cuts the hazuya into the shape and size of the hamon to be worked on each time he polishes the boshi of the hira-zukuri blade. To prevent the hazuya from going over the ji in the kissaki, he uses the thumb of his left hand to block the ji.

Takaiwa is working with a hazuya stone and tojiru paste on the hamon near the kissaki. Because this is a hira-zukuri sword, a narumedai is not used until the very tip of the kissaki is polished.

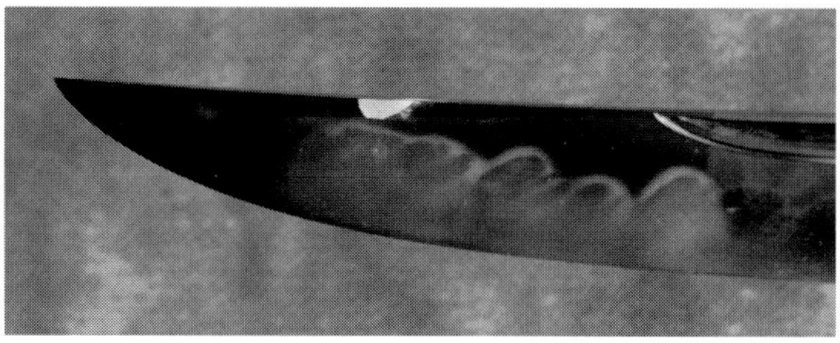

A hira-zukuri blade with a hamon, just below the boshi, that has been polished with a hazuya stone. Takaiwa will next polish the kaeri portion below the boshi, along the mune part of the blade, before polishing the boshi itself, so that he can control the entire shape and appearance of the hamon in the kissaki more easily. The small piece of white hazuya visible on the blade is used to polish the narrow kaeri region. The polished portion of the hamon below the kissaki is white and well defined, while the portion of the boshi that has not yet been polished with the hadori is dark and about the same color as the ji.

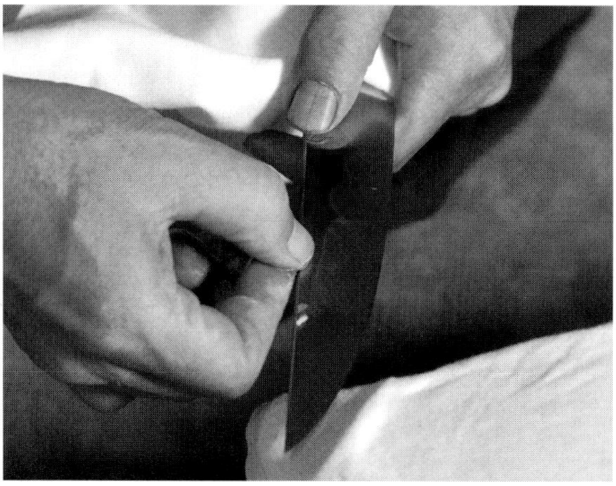

Takaiwa is polishing the *kaeri* (the part of the hamon in the kissaki that runs along and below the mune surface) with the tiny hazuya that is about as wide as the narrow kaeri portion of the hamon.

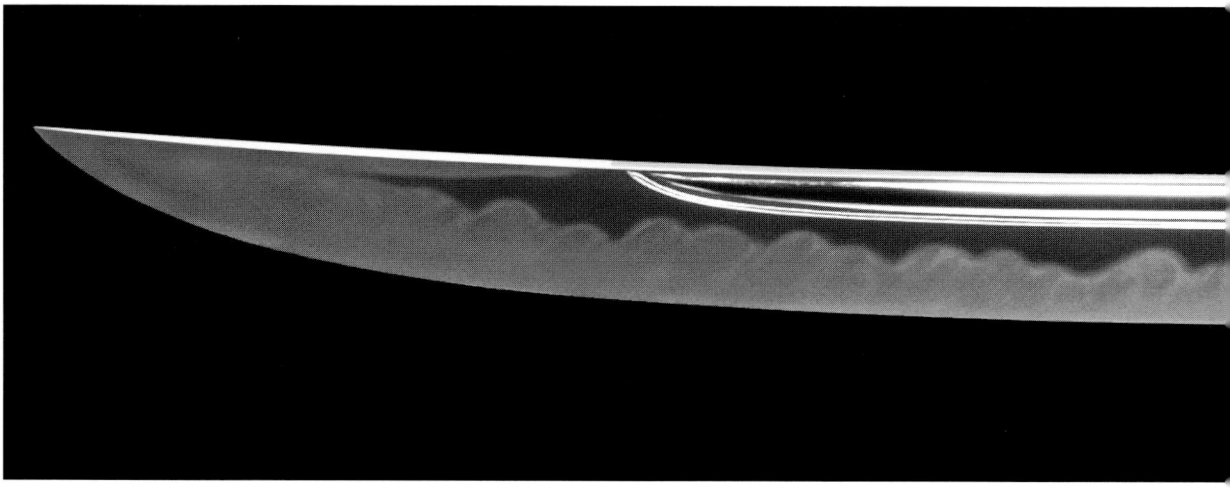

A blade on which the two *nugui-hadori* processes can be compared. The left half shows the completed kanahada nugui/hadori polishing, and the right half shows the results of hadori/sashikomi nugui polishing. The hamon on the left is more stylized, while the one on the right follows the original shape.

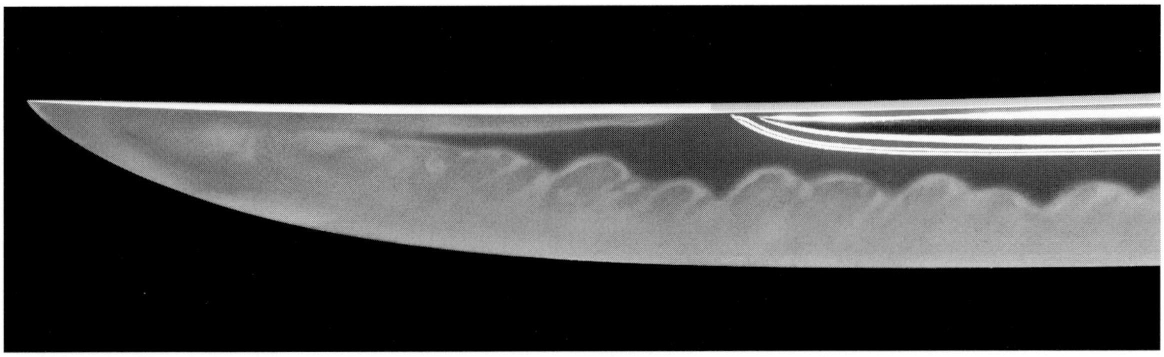

A hira-zukuri blade, the hamon of which has been completely polished with the hazuya.

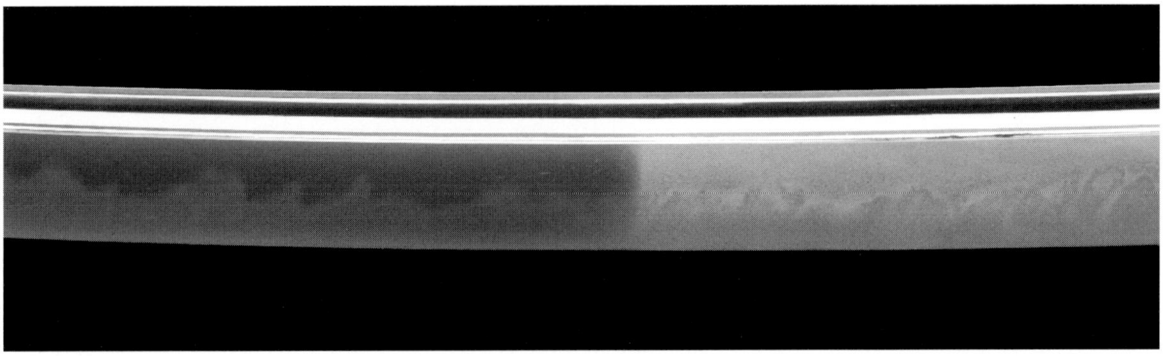

A blade showing the effect of two different polishing processes. The left half of the blade illustrates the effect after kanahada nugui but before hadori, while the right half shows the results after hadori but before sashikomi nugui.

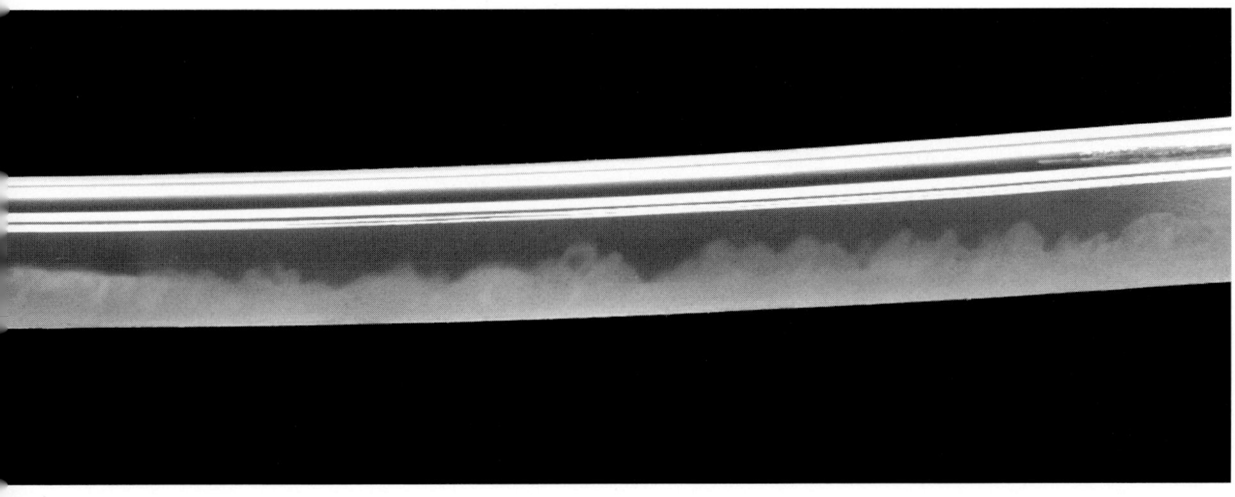

Polishing the Point

MAKING THE YOKOTE LINE

OBJECTIVE

To make the yokote line in the proper location and at the proper angle. The line must be sharp and clear. After finishing with the hadori stage, which whitens and emphasizes the hamon, making the yokote line which defines the kissaki is the next step.

TOOLS

Hazuya, thick plastic or metal tape, bamboo spatula.

Takaiwa's method varies from that of more traditional polishers, but it works well for him. The first step in making the yokote line is to decide exactly where to locate it. If there is a real physical yokote line in place after polishing the ji, the line should be kept where it is. A good polisher will make the yokote line during the early stages of his polishing work with the rougher stones. However, in many cases, the real line has been polished over from an earlier polishing or wear and there might be only a smooth surface where the yokote line should be placed. In either case, whether the line already exists or not, the technique is the same.

The yokote line should be at right angles to the mune where the shinogiji begins to narrow in earnest and to approach the mune surface. Takaiwa places a thick tape on the ji, with the forward edge of the tape in the position where the yokote is to go. The yokote will be defined by

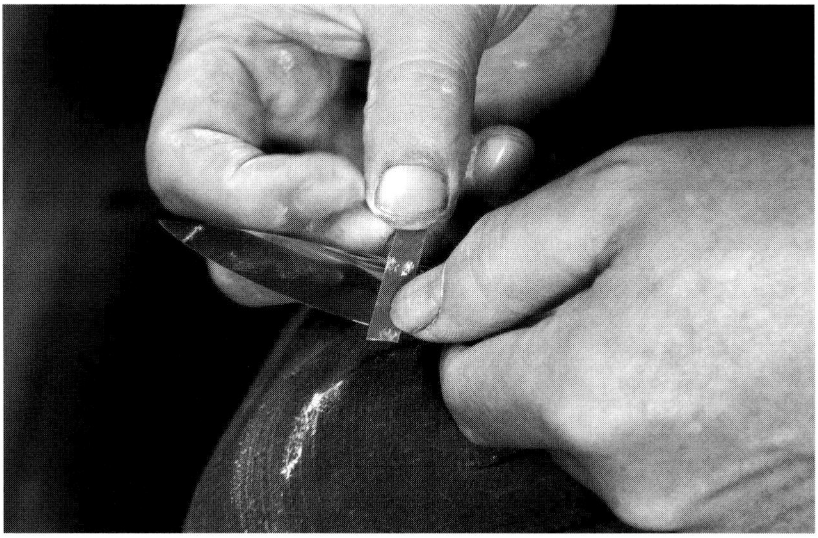

Takaiwa places the thick plastic tape on the ji along the yokote line. In general, he avoids using the traditional bamboo mask.

polishing the kissaki side of the tape and will become visible when the point is polished with a hazuya stone, which will turn the point very light or almost white. In contrast, the ji and hamon will be darker, with the resulting contrast defining the yokote line.

The polisher prepares a hazuya, which is usually a square. However, Takaiwa makes his hazuya trapezoidal (see photograph on page 125) because this shape will leave a gradual transition from almost white to the dark gray color of the unfinished kissaki. At this stage, it is advisable not to have a strong boundary inside the point area between the area polished with the hazuya near the yokote and the remaining

The yokote is made by using a bamboo spatula to move a tiny piece of hazuya stone along the edge of the tape. This arrangement prevents the small hazuya from slipping under the tape and results in a sharp, well-defined yokote line. The tojiru paste on the uchigumori stone (on the left) is used as necessary to lubricate the surface of the point.

Takaiwa grinds the edges of the hazuya stone on an uchigumori stone to smooth the edges before using it to make the yokote line.

unpolished area. The hazuya is moved in a direction parallel to the tape and perpendicular to the lengthwise axis of the sword. The sword is placed on a flat surface and held in place with the thumb of the left hand over the tape. Because the stone is so small, it is moved with a bamboo spatula. First, tojiru liquid is placed on the point area, then the trapezoidal hazuya is used along the tape.

If this step is properly performed, the yokote line should be sharp and clear and the beginning of the point area will be almost white and clearly set off against the rest of the sword.

POLISHING THE REST OF THE POINT

OBJECTIVE

To polish the rest of the point area to match the color and appearance of the area just polished near the yokote.

TOOLS

Narumedai, hazuya.

METHOD

The narumedai is used to polish the point area after the yokote is defined. It is a piece of wood about 1 (or 1½) by 6 by 2 inches (2.5 [or 4] by 15 by 5 cm). The area under the top of the narumedai is cut with a saw four or five times successively to leave a series of slots under the top surface. This leaves it flexible and soft. The top of the narumedai is then covered with about five layers of Yoshino paper, and the whole narumedai with the paper is dipped into a bucket of water so that the layers adhere well to each other. Next, a large piece of hazuya is cut to a width slightly less than that of the Yoshino paper, which is in turn less than that of the narumedai. The hazuya is ground down so it becomes very thin and translucent. This thin hazuya is placed on the paper. The arrangement leaves the polishing surface of the hazuya very flexible and soft. When the point of a sword is moved over it, the hazuya will be able to flex enough to polish the entire point area evenly and uniformly.

Takaiwa removes the plastic tape from the yokote area before using the narumedai. The polisher has to be careful to polish near the yokote but not to go over the yokote into the ji. Some polishers use a strip of Scotch tape to protect the yokote.

The point area is polished by moving the blade up and back over the narumedai in a direction perpendicular to the lengthwise axis of the

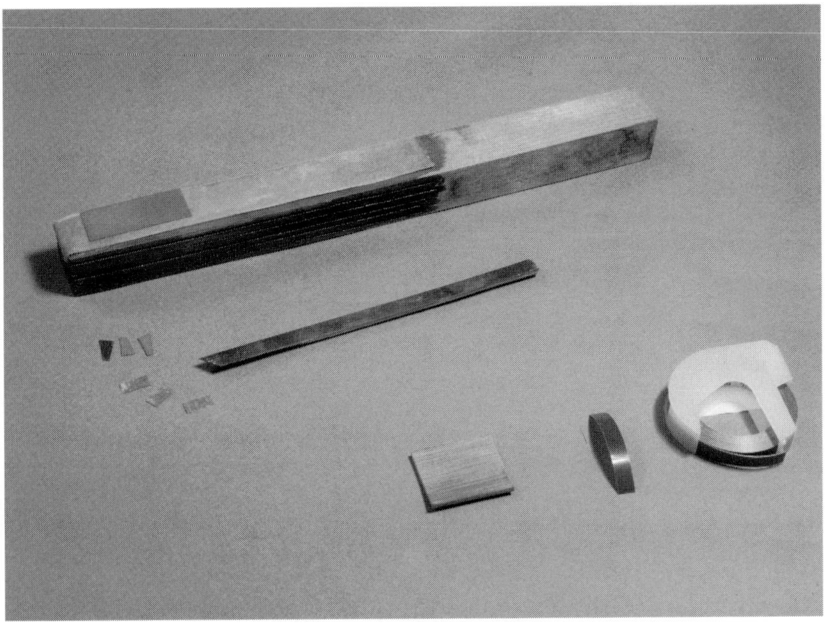

Tools used to polish the kissaki and make the yokote line. The narumedai has about five layers of Yoshino paper on it, and the hazuya is resting on top of the Yoshino paper; small trapezoidal-shaped pieces of hazuya used to make the yokote line; a bamboo spatula to move the hazuya pieces over the kissaki; a square piece of bamboo to define the yokote line and block the hazuya from straying beyond the yokote; a piece of thick plastic tape to mask the yokote line during polishing.

Takaiwa polishes the kissaki with the narumedai, which is held in place with a foot clamp and used in the same position as the large polishing stones. The blade is held so that the left hand holds and directs the blade's movement.

Takaiwa polishes the kissaki on the narumedai. The blade is moved along the length of the stone in small movements. Notice that the sword is completely wrapped in Yoshino paper and then wrapped again in a cotton cloth. The polisher must be very careful to protect the polished sword from accidental scratching.

The kissaki of a hira-zukuri blade is being polished on the narumedai. The very tip of the kissaki is being worked on here. Notice that the tang is elevated above the narumedai.

sword, working from the tip of the point toward the yokote. Only a very small strip is worked on at any one time. The polisher must work carefully and gently to avoid cutting the hazuya on the narume with the edge of the blade.

The area bordering the yokote has to be finished to match the color and finish left by the hazuya used to polish the yokote line. Sometimes polishers are so concerned about going over the yokote at this stage that they leave the area underpolished or nearly so.

Migaki: Burnishing the Shinogiji and Mune

OBJECTIVE

To give the shinogiji and mune surfaces a mirrorlike finish.

TOOLS

Burnishing spatula (migaki-bera), burnishing needle (migaki-bo), powdered horn paste (tsunoko paste), ibota powder (lubricant), cloth.

METHOD

Burnishing the shinogiji. The shinogiji is cleaned with tsunoko paste, then dried and dusted with ibota powder, which is in turn wiped off.

The foundation burnishing is done with a burnishing spatula, and the finish burnishing with a burnishing needle. The polisher starts burnishing near the top of the tang and works his way up the blade, doing on about 1¼ to 1½ inches (3 to 4 cm) at a time. After finishing one side of the shinogiji, the polisher works the other side in the same way.

Before using the burnishing tools, the blade is carefully cleaned with tsunoko paste to remove all dirt and oil. The paste is then removed with water and a clean cloth.

Next, the blade is dusted with ibota powder, which is then wiped off before the burnishing tools are used.

► After having used the burnishing spatula over the entire length of the shinogiji in the foundation burnishing, Takaiwa uses the burnishing needle to remove the marks left by the burnishing spatula during the finishing stage.

In the foundation stage, Takaiwa burnishes the shinogiji with a burnishing spatula in a lengthwise direction.

The results of the foundation burnishing can be seen in the small dark section on the shinogiji.

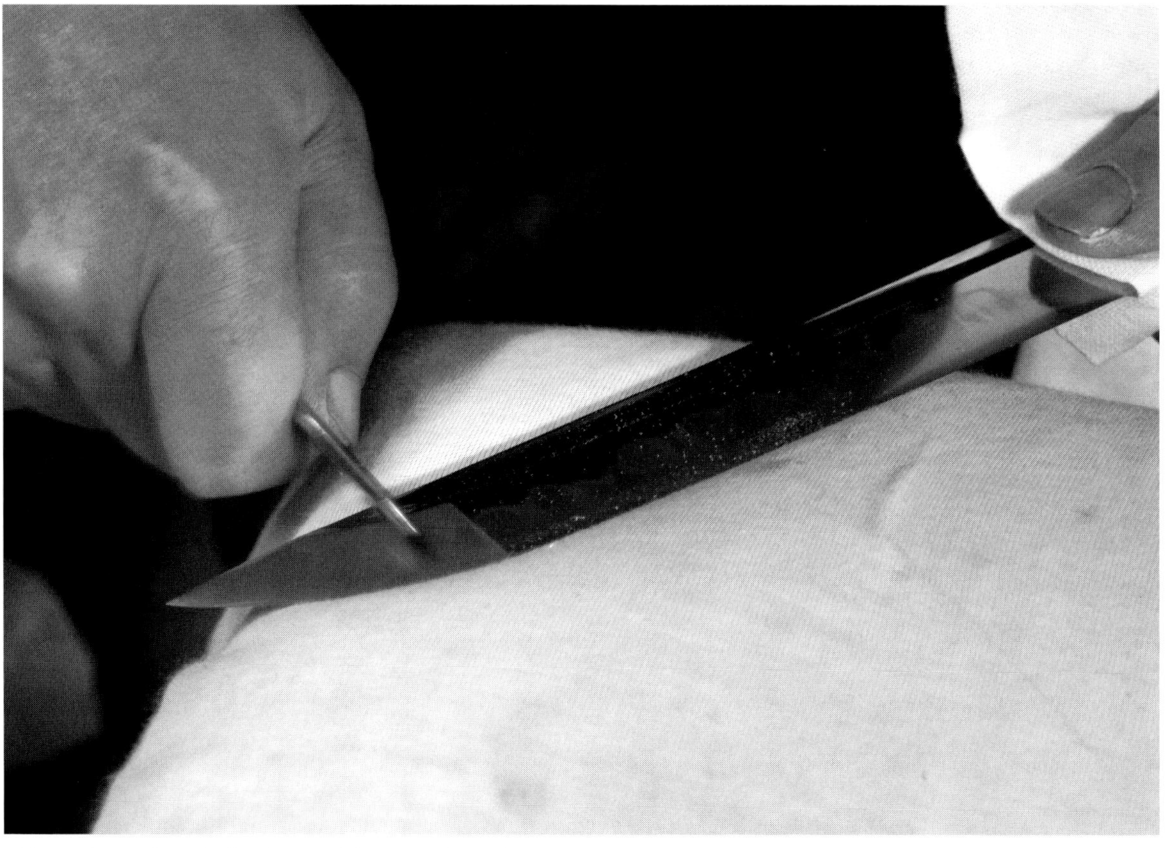

Takaiwa burnishes the ko-shinogi (the part of the shinogi over the kissaki) with the spatula.

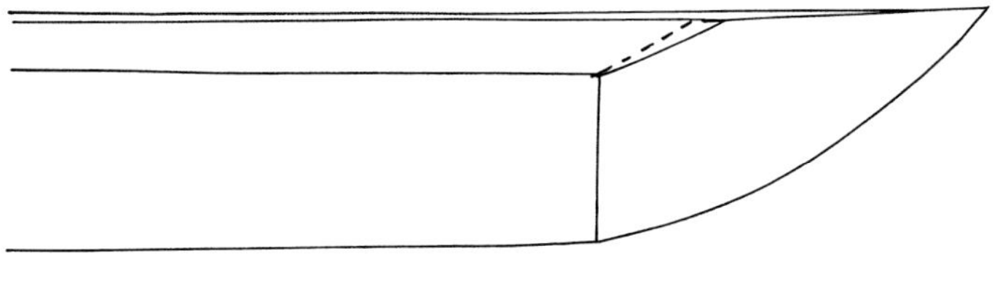

The dotted line indicates a line made by an overrun from the kissaki to the ko-shinogi in narume polishing. This leads to an unfocused ko-shinogi. An overrun can be corrected in the burnishing stage, the two peak lines merged into a single sharp peak.

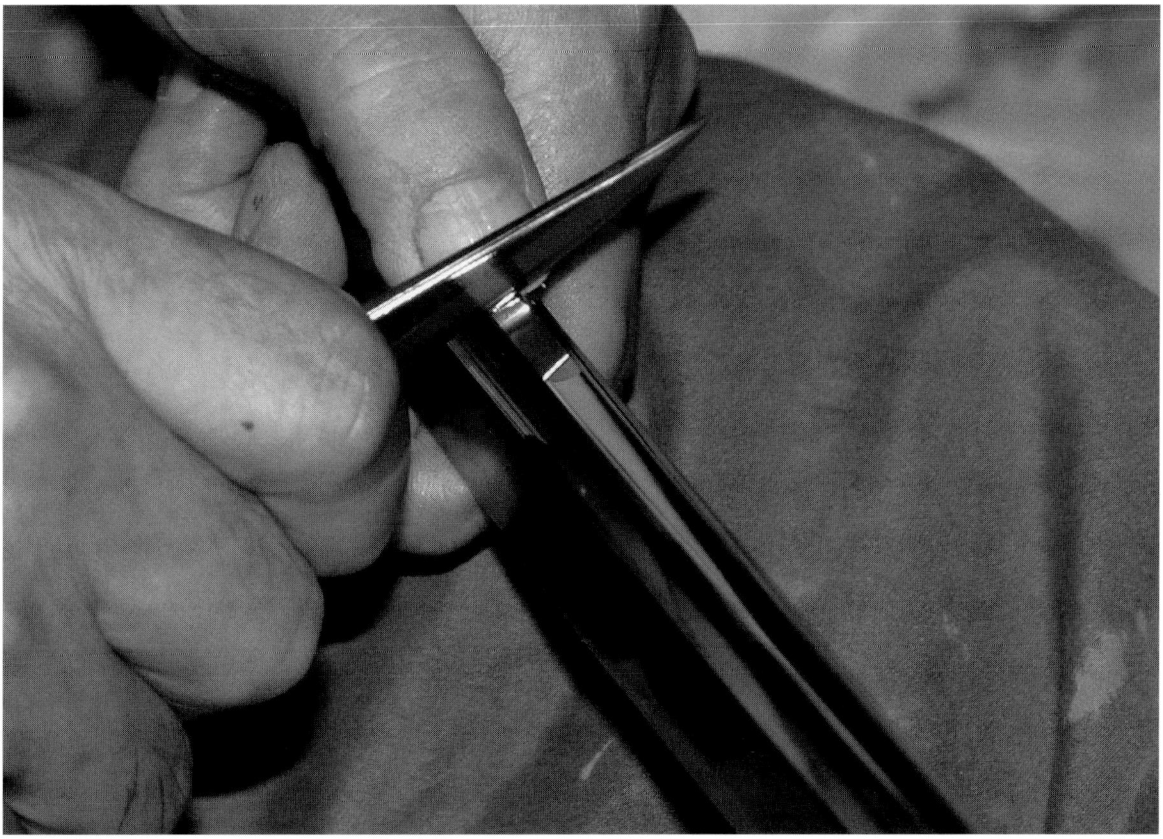

Takaiwa burnishes the mune area with a burnishing spatula.

When beginning to polish the shinogiji near the top of the tang, the polisher uses a bamboo mask (Takaiwa uses tape) that fits into the notch at the top of the tang (the munemachi) and serves as a stop so that the burnishing needle can work right up to the bamboo mask and leave a straight line on the border of the burnished surface. The rest of the burnished surface on the shinogiji is bound by the shinogi and the ko-shinogi. The polisher must be very careful to polish up to the shinogi line and not beyond in order to leave a clean, sharp line. An overrun at the narume stage (where the ko-shinogiji has been inadvertently polished) is often fixed at this stage.

Burnishing the mune: Up to this point the mune has been polished with a synthetic nagura stone, followed by sandpaper. The polisher proceeds to use sandpaper on other parts of the blade.

Before the burnishing, the mune must be cleaned with powdered horn paste. Water is added to the powdered horn to make a thick paste. The paste is rubbed on all of the mune surfaces that will be burnished. The mune is next dried off and dusted with ibota powder (a fine powdered lubricant), which in turn is wiped off.

The first stage of this process is called foundation burnishing. The mune is burnished by rubbing the burnishing spatula over one side of the mune from the base of the blade near the tang toward the area near the yokote. The polisher works on a section about 1 ¼ to 1 ½ inches (3 to 4 cm) long at one time. He begins on the edge near the ji, and rubbing back and forth, works his way to the top of the mune at the ridge of the mune. It is important not to go over the top of the mune and round off or dull the line of the top ridge. The polisher then repeats this process on the other side of the mune and works his way from the base of the blade to the area near the yokote on that side. It is during the burnishing stage that the file marks from the nagura stone are removed.

The second stage of this process is called finish burnishing. The polisher uses a burnishing needle for this stage and goes over the mune's two surfaces in sequence, just as he did with the foundation burnishing, using the ibota to lubricate the process. This stage should remove all of the marks left by the burnishing spatula and should produce a bright, almost mirrorlike finish on the mune. Many of the marks that are to be removed by the burnishing needle were made by the burnishing spatula when it stopped in its back and forth motions during the first part of this process. If the burnishing tools do not move smoothly over the steel, they should be cleaned as described above, while the blade should be cleaned with tsunoko and dusted with ibota again.

The polisher holds the burnishing tools with a cloth so that his fingers or any part of his hand do not rub against the blade as works.

It is best to burnish the side of the mune on the side of the sword furthest from the polisher. The burnishing needle may slip during this process. If the polisher is working on the side of the mune closest to him, a slip could leave a very deep scratch which could mar the surface, requiring a lot of time and effort to remove or repair. By working on the far side, the end of the needle, if it slips, will mostly miss the sword.

Kesho Migaki

OBJECTIVE

Decorative burnishing on the mune and the shinogiji.

TOOL

Burnishing needle.

METHOD

Polishers add decorative burnished lines to the shinogiji under the habaki area and at the top of the unpolished tang area, just below the solidly burnished area. Polishers also burnish a series of three lines on each half of the mune just ahead of the solidly burnished area, and these lines extend up to the area over the yokote, or near the yokote. These lines can be drawn freehand with the burnishing needle in the appropriate areas just before the burnishing work is done on the sword.

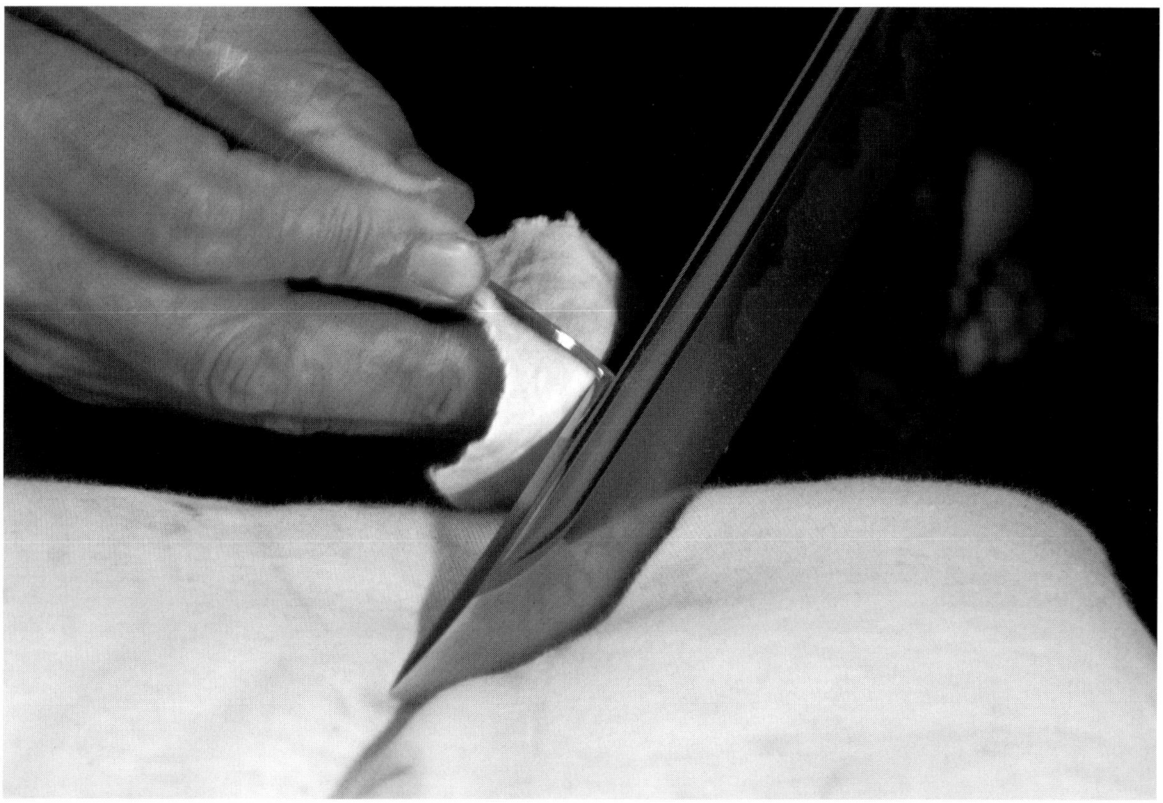

The mune is burnished with the burnishing needle after the decorative burnished lines have been made with the burnishing needle. This second set of decorative lines is made on the whitened part of the mune over the kissaki.

Takaiwa makes decorative burnished lines on the tang.

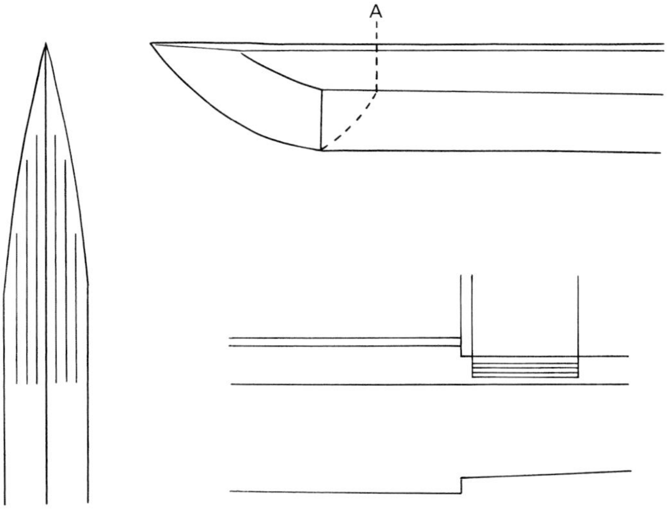

Decorative line on the mune and tang. The six lines on the mune stop around point A.

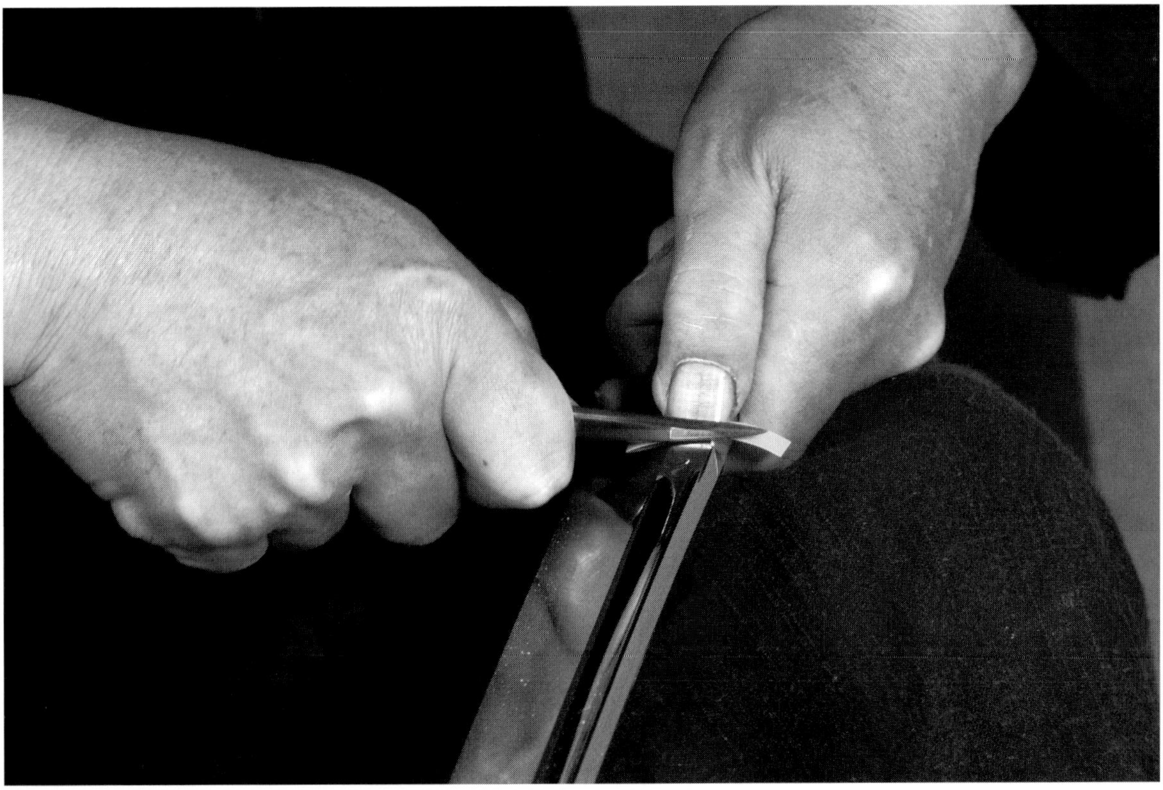

Takaiwa commences burnishing the shinogi-ji, and begins in the area of the munemachi, starting at the top of the decorative lines he has just made. The tape he is holding with his thumb is a mask to define the beginning of the solidly burnished area, and help the burnishing needle make a straight line at the beginning of the burnished area.

Different schools and polishers use different patterns for the lines, and some patterns may actually identify or say something about the polisher. For example, the Hon'ami school of polishing usually alternates thick and thin burnished lines on the shinogiji just above the tang and below the solidly burnished area. A piece of tape is laid over the lines on the tang, and the ends of the lines extending beyond the tape are scraped away with the burnishing spatchula to align them.

The End of the Polisher's Task

If all of the steps have been carried out by a proficient polisher, and he has used good judgment in deciding how to polish the sword at each stage, a modern blade (gendaito) will be finished and all of its inherent character and beauty will be readily apparent. If the blade is old, the same should be true, but the individual hallmarks and characteristics of the blade that indicate which school produced it, and when it was made, should also be apparent.

ADDITIONAL POLISHING TECHNIQUES

The three methods described here are sometimes used on swords to save time.

Tanpan Polishing

OBJECTIVE

To produce a prominent jihada that does not take too much time and effort. The technique is applied when (1) the jihada is weak, (2) the owner of the sword would like to use the sword and does not require an expensive polish, and (3) this technique was used in a previous polish.

MATERIALS

Copper sulfate dissolved in water (copper sulfate:water ratio approximately 1:30).

METHOD

Boiling water is poured on the blade after polishing it up to the uchigumori ha-to stage. Water with copper sulfate is applied to the ji with a Q-tip or cotton wool while carefully avoiding wetting the hamon. The hamon will fade if the water and copper sulfate reach it. Takaiwa uses an uchigumori ji-to stone to polish the ji when the ji changes to a bright copper color. The color changes can take place in as little as ten seconds after the boiling water is poured on the blade. Takaiwa repeats this technique again if the first application does not produce a good result.

Polishing with Nitric Acid

OBJECTIVE

To make the nioi-guchi bright and white. To bring out the nioi-guchi line when it is weak and indistinct after the standard hadori polish.

► Tanpan polishing: the blade is heated by pouring boiling water on it, and a copper sulfate solution is then applied to the ji. This results almost instantly in a bright copper color on the ji.

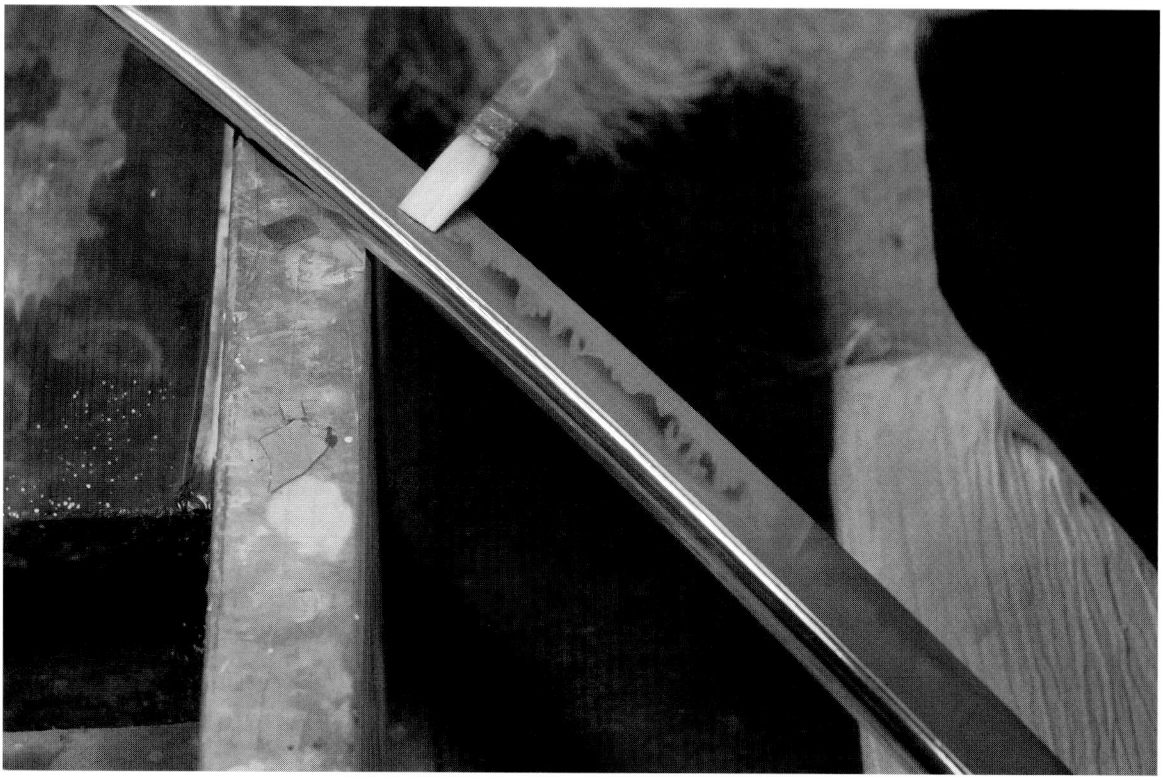

Water with nitric acid is carefully applied to the nioi-guchi and the ha with a brush. The nitric-acid solution must be left on the blade for a very short time. Notice that the steel surface above the nioi-guchi looks dark after this treatment, even as the hamon outline is sharper.

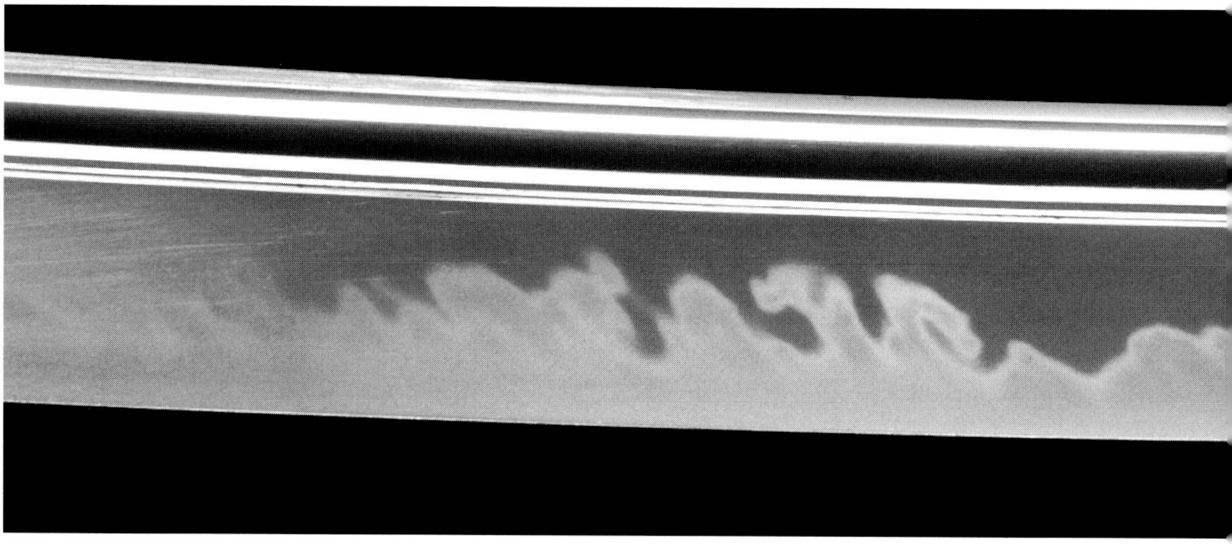

This blade has a prominent nioi-guchi after a combination of nitric acid and sashikomi polishing has been applied.

MATERIALS

Water with nitric acid (nitric acid : water ratio, approximately 1:100).

METHOD

Takaiwa polishes the nioi-guchi area lightly with hazuya stone after using the uchigumori stones. He cleans the blade with a new cotton cloth, so that the blade does not repel the nitric-acid solution. He applies the solution with a brush or cotton Q-tip to the nioi-guchi area and polishes the area with a used hazuya stone (a thin or used piece that has very little stone left on its paper backing) while the blade is still wet. He repeats this process until he obtains the desired result. After this, Takaiwa goes through the usual polishing steps with jizuya, nugui, and hadori. A polisher's goal is to bring out a strong nioi-guchi line to define the hamon. This method does not work with all swords.

Enamel Polishing

OBJECTIVE

To produce a sharply outlined and distinct hamon quickly and easily with hadori stones.

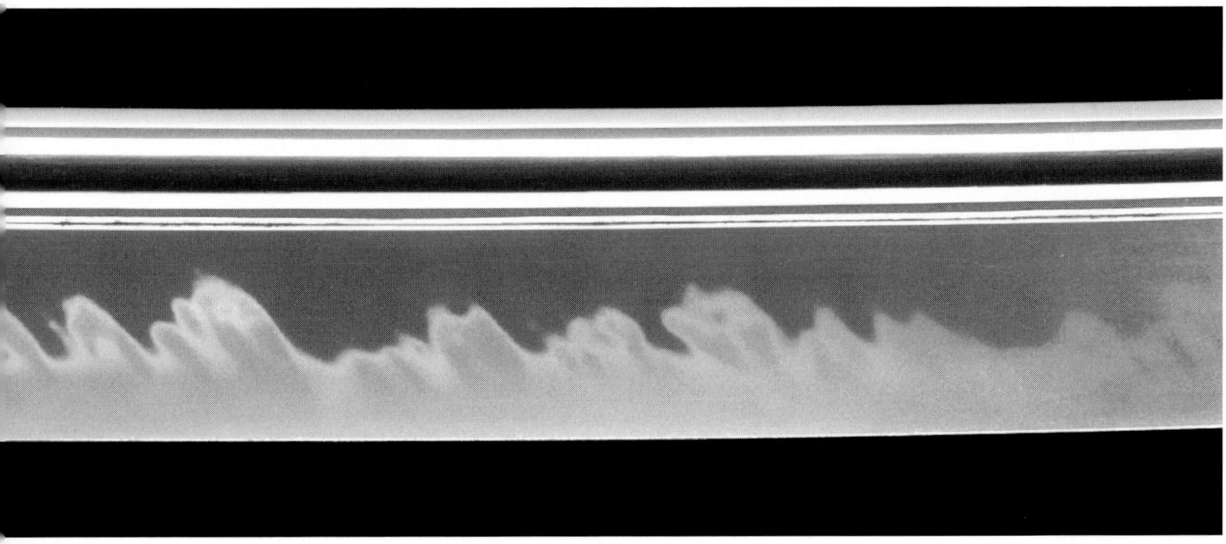

Takaiwa uses enamel on the ji to outline the final shape of the hamon and to mask the ji.

Takaiwa uses a hadori stone with tojiru paste to finish the hamon after the enamel is dry. Since the ji is masked and protected, the hadori polishing can be done easily and quickly.

MATERIAL

Enamel.

METHOD

Paint enamel on the ji to outline the desired hamon shape before the hadori step. When dry, the enamel will act as a mask on the ji, protecting it from even the strongest strokes. The hazuya can only be moved back and forth along the sword in the direction of its lengthwise axis. After this step has been completed, the enamel is removed with lacquer thinner, leaving the shape of the hamon as it was polished with the hadori.

Part III
Profiles of Sword Polishers

The profiles offered in this section are based on interviews with sword polishers that were conducted and written up by Leon and Hiroko Kapp. The first master to tell his tale in these pages is Fujishiro Matsuo, who in 1996 was honored as a Living National Treasure (Ningen Kokuho). Although he passed away in 2004 at the age of 90, the interview, which took place in 1989, is left in the present tense in the interest of clarity.

Fujishiro Matsuo

In Japan today, only two traditions come readily to mind in connection with the subject of sword polishing, namely, the Hon'ami and the Fujishiro traditions. The former is said to date back to the Muromachi period (1392–1573), and the latter to the turn of the twentieth century, the Fujishiro family only having been active in the world of the sword for three generations.

The Fujishiros are well known as polishers, both Matsuo and Okisato being *mukansa*-level (high-ranking swordsmiths) polishers who are also highly respected for their expertise in the areas of sword history and appraisal. One of the standard reference books for collectors is the two-volume *Nihon Toko Jiten* (Who's Who in Japanese Swords), written by Matsuo and Okisato. The family also handles the quarterly *Meito Zukan* (Illustrated Journal of Japanese Swords), was published in turn by Fujishiro Yoshio (before World War Two), Matsuo (postwar to the 1960s), and now Okisato. Both Matsuo and Okisato have contributed photographs to the journal, with Okisato recognized today as one of the country's best sword photographers.

The Fujishiros have made *oshigata*, an official tracing and documentation, of every sword they have polished and Yoshio had collected as a dealer, as well as of swords they examined in their extensive travels over the past fifty to sixty years. The oshigata library compiled by Yoshio,

► Fujishiro Matsuo, 78, is polishing a ji with a jizuya. He is known for his ability to produce beautiful jihada. Interestingly, he often uses his left hand at this stage even though he is right-handed. Note the collection jizuya stones kept in the box in the foreground.

Matsuo, and their brother Takeo continues to grow, due to the efforts of Matsuo and Okisato. In fact, swordmaker Yoshihara Yoshindo believes that the Fujishiro family's oshigata library is now bigger than that of the Tokyo National Museum in Ueno.

In view of the family's prominence in the world of the Japanese sword and the paucity of information on the subject outside Japan, a short overview of the Fujishiros seems in order. After all, Okisato has been featured in the polishing section of the book *Craft of the Japanese Sword*, where his photos illustrate the changes that sword steel undergoes during the polishing process. The following family history is based on conversations we had with Matsuo, his wife, and Okisato.

The family's polishing-related history begins with Matsuo's father Fukutaro, whose three sons—Yoshio, Takeo, and Matsuo—all became involved in swords.

Fukutaro was the first polisher in the Fujishiro line. He was trained under Ishikawa Shuhachi in the late Meiji (1868–1912) and Taisho (1912–26) eras. Of the seven major schools that existed late in the Edo (1600–1868) period, only the Hon'ami school has survived.

Matsuo began polishing with his father in the summers from the age of fourteen. At that time, they lived in the Tokyo district of Kanda, but later moved to the Kudan district, near Yasukuni shrine, where Okisato currently lives. Matsuo, who did not enjoy school, left in the second year of junior high school to become a fulltime polisher. His mother was so angered by this that she moved out of the family home for a while, during which time Matsuo joined his father as a fulltime polisher.

Matsuo was one of seven children, four girls and three boys. He was the youngest son, preceded by Takeo, five years his senior, and Yoshio, twelve years his senior. One of Matsuo's sisters married a maker of *saya* (scabbards) and her son, Takayama Kazuyuki, is a current mukansa-level maker of saya.

Yoshio began as a sword polisher when he was fourteen years old, but later opened a sword shop and thereafter did not do a great deal of sword polishing. This came about because he was accused by an appraiser of selling a fake when in 1930, at the age of twenty-seven, he sold some swords crafted by the swordsmith Minamoto no Kiyomaro (1813–54). At that time, an appraiser's word was considered gospel and people rarely disputed such a pronouncement. However, Yoshio did not agree with his accuser and thus seriously began to study swords and to write in order to validate his assessments.

As a result, Yoshio began to study the more recent *shinshinto* blades, working back to *koto* blades, which was the reverse of the order in which the subjects were generally studied in those days. Yoshio believed that he would be able to apply the results of his study of shinshinto blades to that of the earlier *shinto* and koto swords. Evidence of the swordsmith's work was more prominent in the later blades. Their *mei* (the names of the smiths inscribed on the sword tang) were more legible, the blades themselves were in much better condition (their shapes and hamon still true), and much more historical information was available. A deeper knowledge of these blades would help him extrapolate about earlier blades. And indeed, Yoshio's approach led to new opinions and viewpoints that had not surfaced when the blades were studied in the traditional order.

His initial studies complete, Yoshio, with the assistance of Takeo, published his first book on swords made by the swordsmith Minamoto no Kiyomaro, *Kiyomaro no Mei* (Kiyomaro's Sword Inscriptions), which is also the first book to have been written on the master. Yoshio studied Kiyomaro's life from the perspective of his remaining swords and from oshigata which he made himself. The book is systematically and logically laid out, and explanations of the hamon oshigata are unusually detailed and clear, given that, until then, people had paid scant attention to the finer details of oshigata. The text includes little discussion, since an appraisal of the oshigata (mei and hamon) was thought to suffice, but it does explain that Kiyomaro's name had changed and documents his professional history, all of which information was gleaned from the evidence presented by the swords.

Kiyomaro became famous in Taisho times, probably in part because of this book and the debate it fostered. As a result, Yoshio became well known, started his career as an appraiser and sword historian before the age of thirty, and eventually closed his sword business to puruse appraising and further studies.

Yoshio next wrote a systematic introduction to swords. The handbook contains information such as the names of the various hamon styles and sword-related terminology, swordsmiths' names with assessments of their skills, and genealogies. It was so successful and sold so well that Yoshio was able to make a living exclusively from its sales.

Yoshio also spent a considerable amount of time researching genealogies for other swordsmith families. This he did based on the swords

A portrait of Yoshio at thirty-seven in 1940.

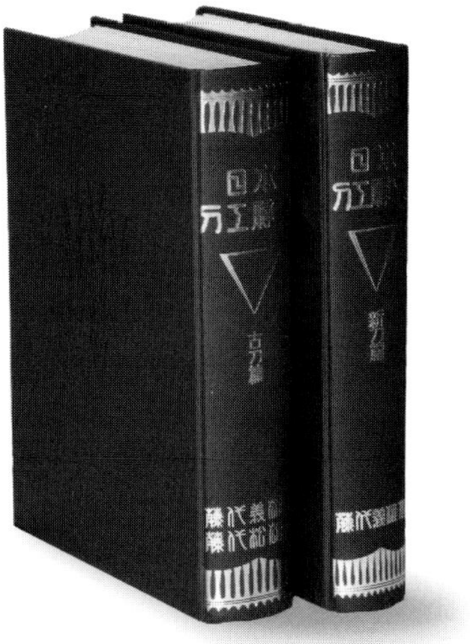

The two-volume *Nihon Toko Jiten* by Yoshio and Matsuo set new standards in the sword world. It is filled with groundbreaking original research and a refreshing candidness. The introductory notes explain the authors' approach as follows: 1. The oshigata introduced in the book were judged by the authors to be authentic works of the named swordsmiths (other opinions were invited). 2. Swordsmiths whose work is extant were given priority over legendary swordsmiths whose work has been lost. 3. Swordsmiths were divided into five ranks from "excellent" to "mediocre" based on the authors' systematic reasoning, as opposed to the ruling convention that follows a hierarchy of status. 4. The lineages of swordsmiths and their schools were reexamined, with new research by the authors offering fresh insights into debatable issues.

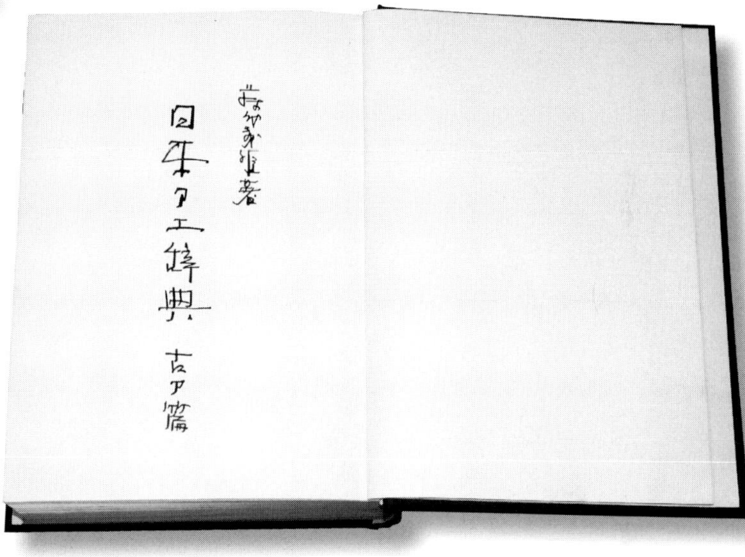

The opening page of the second volume. The title of the book, hand-written by Yoshio, has the elegance of a well-designed typeface.

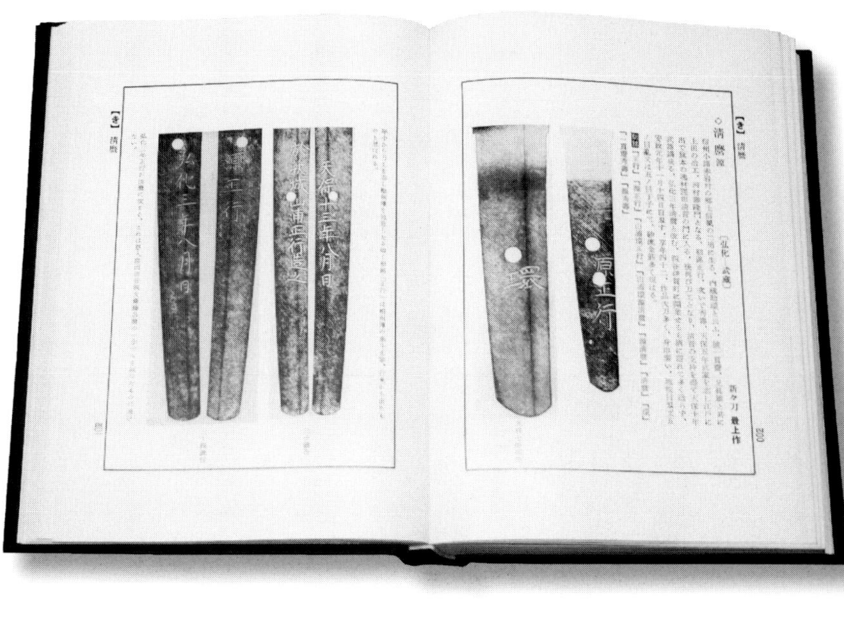

Pages carrying oshigata showing the tang of a sword by the nineteenth-century swordsmith Kiyomaro.

and their mei, relying little on old books and, as a result, he altered many previously accepted genealogies.

One reason that many genealogies were not correct is that, in the Edo period, swordsmith families passed on a swordsmith's given name for a number of generations. As a result many people assumed that the mid-Heian-period (794–1185) smiths had followed the same practice. However, since this was found not to have been the case, Yoshio studied the koto blades and mei in an effort to decipher the genealogies and the relationships between the former swordsmiths. Moreover, because many authors of older books may not have tried to distinguish counterfeit inscriptions (*gi-mei*) from those that were authentic, the old sources often were not accurate.

While Yoshio pursued a career as a historian and critic of swords, Matsuo studied the craft of sword polishing, with his father as his teacher. At the age of fourteen he started his work with finish polishing (shiage) first, and Yoshio soon felt that Matsuo's work was better than that of his father.

Matsuo says that his father was not too careful about his finishing work, although he was very fast, which was the accepted practice in the Taisho era. In addition, Fukutaro was an able interpreter of the marks (*toishi-me*) that stones left on a sword, and had developed an excellent technique for placing the sword on stones.

For a typical polish, Fukutaro charged between 16 yen and 18 yen, which was considered cheap, and if he worked slowly in order to do a better job, he would charge 20 yen. In contrast, a top polisher of the day charged about 50 yen. Fukutaro averaged about one sword per day, a level he maintained even later, when he slowed down a little. Of course, in those days, neither the kind of sword being polished nor what was expected from a polish were the same as they are today: many of the blades then being polished were army swords intended for practical purposes.

In 1937, Matsuo married, bought his own house near that of his father, and became almost independent. The fact that he was already being advertised as a good polisher in Yoshio's sword journal together with the security of having a salary allowed him the luxury of experimentation. He assumed full independence in 1943.

When Matsuo had first started polishing for his father, he had been taught general techniques and methods, but had not been given any absolute rules or formulas. He thus experimented to find what worked best for him and developed his own style. It is quite possible that his skill would not have reached the same heights under Hon'ami tutelage.

One result of his experimentation is that he does *hadori* work starting from the sharpened edge of the blade, rather than from the back (mune) as prescribed by the traditional Hon'ami method.

The Hon'ami school has extremely strict rules concerning all aspects of polishing, and teaches students detailed methods for everything, but Matsuo encourages his students to follow in his footsteps and develop their own polishing techniques.

After World War II ended in 1945, there was no polishing work for two years, while the country recovered. There was a shortage of food and materials, and all weapons, including swords had been confiscated. However, in May 1946, when it again became legal to own a sword, new sword publications appeared.

Matsuo and his family spent the first two years after the war outside Tokyo in order to survive, but Matsuo returned at the end of 1948. He believes that, at the time, only one other polisher besides him was able to make a living in Tokyo, namely, Ono Kokei (1913–98), one of the two polishers who are today Living National Treasures. At the time, the two artisans had to content themselves with polishing primarily for museums, and it cannot have been an easy life. Okisato tells many stories about how poor they were for years after the war. Fukutaro, who also returned to Tokyo and resumed polishing around that time, died in 1949.

It was in 1949 that Matsuo began to attract polishing students because, like most polishers, he found it necessary to supplement the low earnings of his profession.

When the American occupation forces attempted to confiscated all swords at the end of the war, Yoshio was devastated, convinced that this marked the end of the Japanese sword, around which he had built his whole life. A few months later, at the age of 43, Yoshio simply disappeared and was never heard from again. As of this writing, his fate remains unknown.

In 1940, Yoshio had started publishing the quarterly sword journal, the *Meito Zukan*, but eventually had to stop because of the war. In 1951, Matsuo felt that he should restart the publication in his brother's memory. However, he lacked the financial means to print and distribute the magazine, let alone photocopy it. So, for his first issues he used sunlight to print it on blueprint paper and displayed the single copy behind the glass windows of his shop in Tokyo. His wife did the writing and printing. Once he had got this magazine off the ground, he began to do more writing, appraisals, and oshigata to keep it going, and it remains one of Matsuo and Okisato's major endeavors.

Matsuo finds it most enjoyable to polish different kinds of swords in an effort to make them look good, and finds that koto blades produce rather variable results and depend greatly on the individual polisher and his methods. He considers it a challenge and fun to do a good job and make whatever he is working on look like a good sword.

One particular incident in 1926, early in his career, helped establish his reputation. Someone had had a sword by master swordsmith Kunitoshi (mid-13th–early 14th century) polished by an expensive polisher who had charged 50 yen. However, the polish was not satisfactory and the nioi line and hamon were not visible. On seeing this, Yoshio took the sword for repolishing to Matsuo, who was twenty-three at the time. Matsuo polished it well, bringing out the nioi and full hamon, and charged about 20 yen for the work. This made him very well known and from then on, he received many good swords on which to work.

Polishing history is rather vague and difficult to document, but Matsuo does have his opinions on the subject. He feels that although polishers were already using hazuya before the Keicho era (1596–1614), polishing only had allowed the hamon—and probably not the jigane or jihada—to be visible. He believes that the jizuya and nugui polishing steps came into use during the Edo period, and that the burnishing and hadori steps were only developed well into Meiji times. Matsuo believes that the value of swords as an art form was most important during the Meiji period, as a result of which polishing and appearance became much more important, as did the buying and selling of swords. Before that, he believes there had been little buying and selling of swords, so the appearance of a sword played a lesser role. He remembers seeing good swords when he was young, although they had no decorative burnishing (*kesho migaki*) at that time.

Matsuo believes that a good sword does not need a hadori finish, that the ji and ha should show up clearly, and that all of the sword's character and beauty should be apparent. His favorite swords are those made by the swordsmiths Ko-Bizen Masatsune (late 10th–early 11th century), Nagamitsu (late 13th–early 14th century), and Nagasone Kotetsu (ca. 1604–78).

Fujishiro Okisato

Fujishiro Okisato (b. 1947) is the third generation in his family to be a sword polisher and a contemporary of Takaiwa Setsuo. The Fujishiro school is very new by Japanese standards, having been established by his father. Before the Showa period (1926–89) there were seven different polishing traditions, but today there remain only two: the Fujishiro and the Hon'ami.

Okisato's father, Matsuo (1914–2004), began polishing swords before World War II and learned his skills from his father, Fukutaro, who kept his rates comparatively low and was a quick worker, usually polishing a sword in one day. Matsuo became a prominent polisher whose work differed enough from that of traditional polishers to be considered a new tradition. As a polisher, Matsuo was honored as a Living National Treasure, and his brother, Yoshio, was one of the most prominent sword appraisers in Japan in the period before the end of the war.

In high school, Okisato was interested in astronomy and thought about going to college to study science. However, his father told him that if he wanted to polish, he should begin immediately after high school and forego college. Okisato decided to become a polisher and began working part-time with his father in the third year of high school, becoming a full-time sword-polishing student after graduating

from school. Surprisingly, before his senior year in high school, Okisato had never even looked at swords.

Okisato says that modern swords require more time and effort to polish, but he works on many of them now (though it should be noted that he has worked on numerous koto and shinto swords as well). He finds them interesting because the smith often does not know exactly what their final form will be, or even should be, while he, as polisher, is the first one to know exactly how the sword will turn out. A new sword can take at least twice as long to polish as an old one. Although at the start, a new sword is polished with very rough stones, a process that requires a lot of time and effort, Okisato enjoys the work. He also notes that the quality of new swords is improving, the steel being less brittle and the blades better tempered than in the past.

Bearing in mind that it is important for a polisher to emphasize the best points of the sword, Okisato believes that the *yokote* area is one of the most critical parts of the blade, which must also display a good balance and clear differentiation between the ji and the hamon. Sometimes, however, when polishing an older sword he finds he cannot do much to restore the shape, since to do so would require the removal of so much metal that the sword's aesthetic balance would be destroyed.

◄ Fujishiro Okisato is polishing a long *tachi* sword made by Osafune Norimitsu in 1447. The sword is 12½ feet long (3.8m), the body being 7½ feet (2.3 m) and the nakago 5 feet (150 m). It weights about 29 pounds (13 kg). The sword is owned by Kibitsu shrine in Okayama Prefecture. The blade was covered with rust before Okisato polished it in 1992.

Yanagawa Seiji

Yanagawa Seiji (b. 1948), who became a mukansa polisher in 1995, lives in Sakura, just north of Tokyo and very close to Narita Airport. Yanagawa was attracted to swords from the age of eighteen and wanted to make a living in a way somehow involving them. Since he knew little about swords at the time, he asked the local police department where he could see swords and learn more about them. Eventually, he obtained an introduction to a sword collector's association and met the polisher Ono Kokei, who was later to become a Living National Treasure.

Ono suggested that polishing might be the best sword-related activity in which to make a living. In the beginning, Yanagawa wanted to become a student of Fujishiro Matsuo but, when it became clear that there would be no openings in Fujishiro's workshop for a long time to come, he apprenticed himself to Ono in 1968 and spent ten years working and living in the Ono household.

Yanagawa spent the first three months of his apprenticeship watching the other students polish, learning how to hold a sword, as well as cleaning and polishing the grooves in swords. He began by working with modern swords which, at the time, were poorly made. Excessively thick and poorly shaped, they were perfect as swords on which new students might cut their teeth. In time, Yanagawa performed the many

Yanagawa Seiji is polishing a sword with a jizuya. Various stones rest on the shelves. The photograph on the wall next to the small shrine is Yanagawa's master, Ono Kokei.

other tasks that were usually given to new students: making hazuya and jizuya stones; grinding the kanahada to make nugui; and correcting and finishing the shapes of new swords.

At the time of his apprenticeship, students typically remained with their teachers for about ten years. Yanagawa was a relatively able polisher after about three years, but chose to stay the full ten. During the remaining period, he received some payment (about one-third the salary of an office worker of the day) for his work and became an independent polisher in 1978.

Yanagawa says that, while some old swords are in very good condition, others are seemingly worn out after having been polished too many times over the centuries. A worn out and shortened short sword (*suriage*) is often very difficult to polish and restore to its original condition. Certainly, the better the sword, the better the job a polisher can do. He also notes that the best swords from each of the old swordmaking schools are the acknowledged standard for what is expected, and that polishers today are required to meet these standards.

Polishing a modern sword does not necessarily take much more time than if the sword were old. Yanagawa polishes a lot of windows (a small section polished in a new blade to check the appearance of the steel and hamon) as a favor for young swordsmiths. He notes that many young smiths these days lack experience in creating the final shape of a sword, as a result of which they will sometimes turn out a rough sword and ask a polisher to produce the final shape.

If a young smith has no image of the final shape he wishes to produce, it is left to the polisher to finish the sword. While many smiths use old swords as models, it is imperative that modern smiths develop their own shape and style, because the individuality and character of a sword is so important. Some of the problems Yanagawa encounters in connection with new swords include blades that are too thick in the center and have a round cross-sectional shape; those that have an excessively steep or high back ridge; and ones with a poor or rough shape overall.

Yanagawa works mostly on old swords, in the belief that the jigane and jihada are harder to enhance in modern swords than older swords. This might have something to do with the forging techniques, although some swords from the Kamakura period (1185–1333) that are National Treasures share some of the features of good modern swords.

In an interesting side note, Yanagawa makes sword guards (*tsuba*) as a hobby and has won prizes for these creations. His tsuba feature a very fine grain or dot background on a black metal surface, and he uses such decorative themes as grain and plant motifs.

Shinozaki Masanori

Shinozaki Masanori (b. 1950) developed an interest in sword polishing at a young age, after realizing that a neighbor was a sword polisher, although at the time he did not understand the role of a polisher in the overall process of swordmaking. His interest was further piqued by the fact that his father kept two swords at home, and Shinozaki remembers when one of them was polished. So it was at the age of seventeen, in his second year at high school, that Shinozaki bought a sword book by Iida Kazuo, a respected author in the field.

Thereafter, Shinozaki looked at the family swords in a new light, realizing that the neglected weapons needed polishing to restore them. He sought to learn more about polishing and polishers, and not knowing how else to proceed, eventually called Iida for advice. Iida introduced him to a major figure in the sword world, Sato Kanzan, whom he visited at the Nippon Bijutsu Token Hozon Kyokai (NBTHK; The Society for the Preservation of Japanese Art Swords). Sato suggested that Shinozaki should enter the polishing school run by Nagayama Kokan. However, since Shinozaki was unable to afford the tuition fees, Kanzan then suggested he become an apprentice of Ono Kokei.

On a visit to the NBTHK, Shinozaki watched Ono polish a sword. Because he had just taken on a new student, Yanagawa Seiji, Ono suggested that if Shinozaki wished to be a polisher, he should meet Onishi

Tokiji, who was one of Ono's best students. Eventually, after finishing high school, Shinozaki became Onishi's apprentice. However, after only two years of training, a rift developed between master and apprentice, and Shinozaki left to set up on his own.

Shinozaki says that one of the best things that happened to him was having his fellow student, Sato Nobuaki, introduce him to the swordmaker Yoshihara Yoshindo. Yoshihara, about twenty-six or twenty-seven years old at the time, had been working independently for five or six years and had already won several prizes in the annual NBTHK competition.

Shinozaki polished a number of swords for Yoshihara and many of them won prizes at the NBTHK event. Shinozaki's commissions grew as a result.

When polishing a new sword for the first time, Shinozaki believes that one of the most important steps in the foundation polishing stage is the use of the binsui stone. It should be neither too rough nor too fine to bring out the basic form of the sword. However, Shinozaki employs the binsui in a manner quite different from that prescribed by both the Hon'ami and Fujishiro schools.

Whereas the two schools require that the sword be moved crosswise over the binsui stone and travel only over a short area of the stone (about 3 inches/7 to 8 cm), Shinozaki polishes a sword using the full length of the stone. He learned this method from his teacher Onishi, who learned it while an apprentice before World War II from Hirashima Shimazo. The advantage of this method is that it shortens the amount of time required to polish a sword. The disadvantage is that the ji, where the steel is soft compared to the ha, wears down rather quickly and, as a result, a height discrepancy on the surface can develop between the ji and the ha.

Shinozaki's ideas regarding the other stages of the polishing process are also at variance with those prescribed by the more traditional schools. When using the hadori stone, for example, members of the Hon'ami school customarily have the sword's cutting edge facing away from the polisher and work from the sword's unsharpened back ridge. Members of the Fujishiro school, meanwhile, work with the cutting edge facing the polisher and begin work with the cutting edge. Shinozaki believes that more force can be applied by the thumb in the Hon'ami method.

Shinozaki Masanori is preparing to burnish the shinogiji. The shoji sliding paper window softens the sunlight, which makes the workshop comfortable for polishing.

When it comes to the hadori, Shinozaki says there is another difference between the two schools. The Fujishiro method requires that the individual waves of the hamon be worked on one by one, while the Hon'ami tradition requires a polisher to treat two or three waves as one and to work on this area accordingly, with the end result that each grouping becomes one wave once the final hadori stage has been completed.

When asked to define a good blade, Shinozaki mentions the need for an excellent jigane, explaining that the color should be neither too dark nor too bright; the pattern in the steel (jihada) should be good;

and that the steel should display a host of features (*hataraki*) that originated during forging.

Shinozaki admires the work of the renowned swordsmith Ko-Bizen Masatsune (late 10th–early 11th century). Many of Masatsune's swords remain, and all have a beautiful *itame-hada* and *mokume-hada*, with a very good utsuri. Of all the koto swords, Shinozaki is especially fascinated by the workmanship of the Aoe school. He enjoys the challenge of bringing out the unique beauty of the jihada and the hamon. Aoe blades are well known for their distinctive *chirimen-hada*, or "crepe jihada." Excellent pieces in this school have a subdued shine to the steel. He explains that a hard jizuya will bring out the distinctiveness of the chirimen-hada, but it will make the ji too shiny and sharp, which will spoil the beauty of Aoe swords. A jizuya stone that is too soft, however, produces a subdued shine on the blade, but misses a distinctive chirimen-hada. He enjoys selecting the best jizuya for Aoe swords and developing better polishing methods for them.

Among shinto swords, Shimazaki likes the work of the Nanki Shigekuni school. He finds the balance between ji and ha excellent, and the bright and clear hamon extremely beautiful. In the shinshinto school, Shimazaki feels that no smith can complete with Minamoto no Kiyomaro, whose swords are an embodiment of "the spirit of the samurai." Beautiful and virtually perfect in terms of shape and proportion, Shimazaki cannot envisage swords with better-crafted ji or hamon. That said, Shimazaki nevertheless believes that the jizuya and the nugui techniques of the Fujishiro school bring out the craftsmanship of Kiyomaro swords better than do the techniques of the Hon'ami school.

Shinozaki became a mukansa in 1999. Although his career really took off when he began to polish modern swords, he says they are much harder to polish than old swords and require more time and physical exertion. Interestingly, although he began under the tutelage of Yoshihara Yoshindo, he has polished all the swords made by Yoshindo's son, Yoshikazu, for the modern-sword (gendaito) contest over the past fifteen years or so.

Takaiwa Setsuo

Takaiwa Setsuo (b. 1947) is a second-generation polisher whose father, originally from Hokkaido, moved to Izumo to be near the newly active smelter (*tatara*), which was being reactivated to make *tama hagane* steel for the Yasukuni Shrine swordsmiths around 1940. His task was to train new polishers, as well as to polish gunto, the modern Japanese swords then being made for the Japanese military.

After the war, when possessing a sword was proscribed for a brief period by the occupying U.S. forces and there was, thus, no work for polishers, Takaiwa's father became a salaried worker and polished swords as a hobby. However, with the economic upswing in 1963 there arose a renewed interest in swords, and his father resumed polishing swords full time.

Takaiwa himself recognized that sword-polishing offered better remuneration than the work of a salaried employee, and decided to become a sword polisher at the age of sixteen. He became a student of Fujishiro Matsuo, who had been poor and lacked work during the immediate post-war period. Takaiwa lived in the house as an apprentice and remembers having to spend about half an hour every morning cleaning the house and taking care of household chores. Even so, he spent thirteen years with his teacher.

Fujishiro was a very good polisher, and so Takaiwa, not sure that he

◀ Takaiwa is examining the condition of a sword before he starts polishing. The stone on the wooden block is koma-nagura.

would ever be able to match his teacher's skill and become an independent polisher, remained a student for a long time. However, once he left Fujishiro's shop in around 1976 at the age of 28, he had enough work to easily support himself.

At this time, sword dealers and shops noticed that the swords which sold well were those with a strong contrast between ji and hamon, which was produced by the polishing style employed by Fujishiro. His style produced a dark ji and a white or strongly contrasting hamon, with a strong *habuchi* between them. This, Takaiwa believes, is Fujishiro's contribution to the art of polishing. Previously, many polishes had had a cloudy, uninteresting ji, a finish that made it hard to see and, thus, hard to appreciate the subtle features of a sword.

Following the technique of his teacher, Takaiwa tries to polish swords so that the ji and its details are evident in the light generally available in any household, and do not require the stronger, focused lighting set up specifically for sword-viewing purposes.

When developing his own polishing techniques, Takaiwa noticed that the ji on the blades of koto swords looked dark and rich with details of the jihada and the ji clearly visible, while the ji on shinto blades tended to be bright and reflective. He wanted shinto blades to look more like the koto blades.

The standard polishing methods of the time did not, however, permit him to explore this line of thought so that, in an effort to improve his polishing techniques, he sought out older stones that had been quarried in the Edo period to see if the results produced might not be noticeably different. Takaiwa notes that, although synthetic stones have been available over the last ten to fifteen years, natural stones were used for all polishing stages when he was a student.

Today, the synthetic stones are generally used up to the nagura stage, and natural stones for the last stages. In the early stages, synthetic stones work well to correct the shape, sharpen the sword, and remove damage. However, from the nagura to the uchigumori stages, natural stones should be used if the polisher wishes to bring out the detail of the ji and hamon. Takaiwa looked for older natural stones and found them in polishers' shops and in the warehouses of merchants who dealt in natural stones, and now has a supply sufficient to last the rest of his career.

According to old historical documents, these old stones were very expensive. They are certainly rougher to the touch than their modern counterparts and seem to grab the steel surface better, probably due to their different mineral compositions and the locations of older quarries.

When buying newly quarried stones, Takaiwa tests them and says that about one-fifth meet his expectations.

Takaiwa, who has never had an apprentice, provides advice and suggestions to a number of polishers.

He can polish a modern sword in about ten days, but requires only about half as long to polish an old blade. He believes it is key to know precisely when to stop working with a particular stone and move on to the next stage.

As an apprentice, he never saw Fujishiro scold a student or dictate what a student should do. Instead, he would point to some part of the sword and say, "Here's a problem, so do something about this," an indirect approach that forced students to think for themselves. As a result of his training environment, Takaiwa began thinking a lot about the stones themselves. Fujishiro also preached that every sword has its own character and nature, and the polisher should let this guide him.

Today, Takaiwa feels that all professional polishers are working at almost equal, highly professional levels, and that it is difficult to compare their work. Moreover, polishing is also an evolving art, with continuous improvements and innovations in techniques and materials.

Takaiwa recalls his apprenticeship fondly, even though the schedule could be grueling. He was expected to work just about every day and only took off two days a month; in exchange, he received room and board. The hours were long, typically from early afternoon to midnight or later. One of his fellow students, Sumi Kenzo, used to make a point of going to the public baths daily just to get out of the house.

One of his most memorable recollections is when he was sixteen years old, soon after he began his apprenticeship. Takaiwa had just finished polishing a short sword with an uchigumori stone. As he was taking it to his master for inspection, the blade came loose from its scabbard. Without a moment's hesitation, Fujishiro reached out and grabbed the blade with his bare hand, heedless of the injury the sharpened blade could cause, more concerned about the almost certain damage to the blade had it struck the ground. His reaction was so swift that it startled the young Takaiwa.

Later, Takaiwa learned the rest of the story. Fujishiro believed a blade should be protected at all costs, it being a swordsmith's precious creation. His reverence for the sword was so firmly rooted in his being that, when Takaiwa's blade had started to drop, he had risked severely injuring his hand and fingers without giving the matter a second thought. Luckily, he had hurt himself only slightly. But the incident left an ever-lasting impression on Takaiwa.

Appendices

HISTORICAL PERIODS

Yamato period (大和時代)	645–710
Nara period (奈良時代)	710–794
Heian period (平安時代)	794–1185
Kamakura period (鎌倉時代)	1185–1333
Nanbokucho period (南北朝時代)	1333–1392
Muromachi period (室町時代)	1392–1573
Azuchi-Momoyama period (安土桃山時代)	1573–1600
Edo period (江戸時代)	1600–1867
Meiji era (明治時代)	1868–1912
Taisho era (大正時代)	1912–1926
Showa era (昭和時代)	1926–1989
Heisei (平成)	1989–
* Sengoku period (Warring States period)	1467–1568
* Bakumatsu (end of the Edo period)	mid- to late 19th century

MAIN SWORD CATEGORIES

Jokoto—Yamato period to Mid-Heian period
Koto—Mid-Heian period to 1596
Shinto—1596 to 1781 (Tenmei era)
Shinshinto—1781 to 1876 (the year the *Haitorei* was issued)
Gendaito—1876 to the present

KOTO AND THE FIVE TRADITIONS (*GOKADEN*)

The term *Gokaden* is one of the most basic words used to describe various types of swords not only by connoisseurs but also by those with a general interest in the Japanese sword. The *Gokaden*, which means "five traditions," refers to the five distinct styles of workmanship developed in the provinces of Yamato, Yamashiro, Bizen, Soshu (or Sagami), and Mino, during *Koto* times. Later the word came to be used to describe swords' types in general with no association with particular regions.

Typical workmanship of the *Gokaden*

Yamashiro tradition

Sugata: *Torii-zori*, smaller *mihaba*, slightly bigger *kasane*, *funbari*, and *ko-kissaki*.

Jihada: Dense *ko-itame-hada* or *ko-itame-hada*.

Hamon: Basically *suguha* in nie or *ko-nie deki*.

Yamato traditon

Sugata: Deep *torii-zori* with high *shinogi* and slightly extended *kissaki*.

Jihada: Mostly *masame-hada*.

Hamon: *Suguha* with rough *nie*.

Bizen tradition

Sugata: *Koshi-zori*, with standard *mihaba* and bigger *kasane* with *chu-kissaki*.

Jigane: *Itame-hada* often accompanied by *utsuri*.

Hamon: *Choji midare* in *nioi deki*.

Soshu tradition

Sugata: Shallow *torii-zori*, bigger *mihaba* and smaller *kasane*. *Chu-kissaki* and *o-kissaki*.

Jigane: *Itame-hada* with *ji-nie* and *chikei*.

Hamon: *Gunome*, *midare-ba* and *hitatsura*. *Nie*, *sunagashi*, and *kinsuji* are often seen in the *hamon*.

Mino tradition

Sugata: Shallow *torii-zori* and *saki-zori* with high *shinogi*. The size of *moto-haba* and *saki-haba* is almost the same, and the *kissaki* is slightly extended.

Jigane: Dense *itame-hada*.

Hamon: *Gunome* in *nioi deki*, *gunome midare*, and *hiro-suguha*. *Togari-ba*, or lines resembling spikes, appear in the three types of *hamon*.

Each tradition is defined by the workmanship produced when the style reached a certain level of stability and completeness. The workmanship was subject to the trends of the times. For example, each tradition had a much longer blade with a bigger *mihaba* and smaller *kasane* and *o-kissaki* in the Nanbokucho period. However, in the Sengoku period, when *tachi* were being replaced with *katana*, the size of the blade in each tradition became much shorter.

TYPES OF SWORDS

Tachi: A long sword which is worn with the cutting edge down and is suspended from a belt.

Katana: A long sword which is worn with the cutting edge up and is supported by an *obi*, or sash.

Wakizashi: A short sword, generally with a length between 12 and 24 inches (one to two *shaku*).

Tanto: A dagger with a maximum length of approximately 12 inches (one *shaku*).

Ken: An old style of straight blade with a double edge. This can be in any length from a *tanto* (12 inches) to a sword (over 24 inches).

Naginata: A polearm with a blade mounted on a very long pole. *Naginata* have a characteristic shape with a wide, strongly curved point.

Yari: A polearm which has a straight blade, and which sometimes has cross arms. These come in a variety of shapes, from simple straight blades to various types of cross-shaped blades.

Hoko: An older type of *yari* which has a socket for the pole to fit into (instead of the usual tang which fits into the top of a pole), But some *hoko* have traditional Japanese tangs.

TACHI.

KATANA.

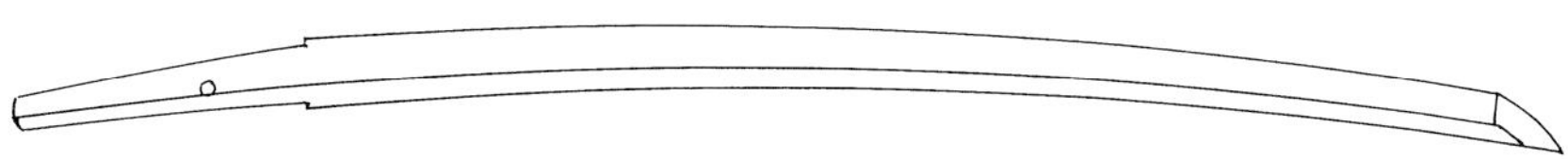

WAKIZASHI.

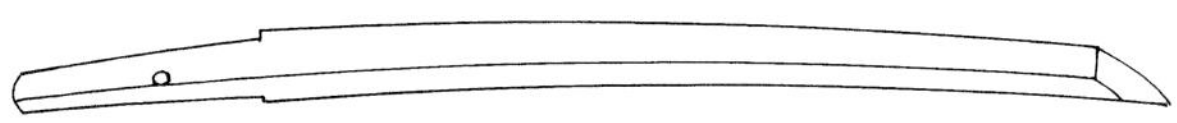

TANTO.

KEN.

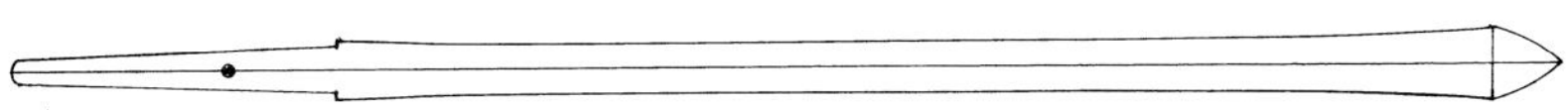

KEN.

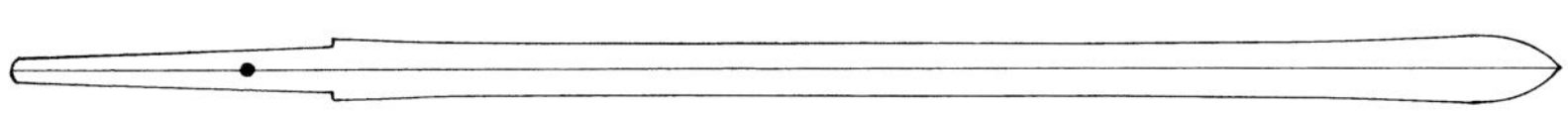

NAGINATA.

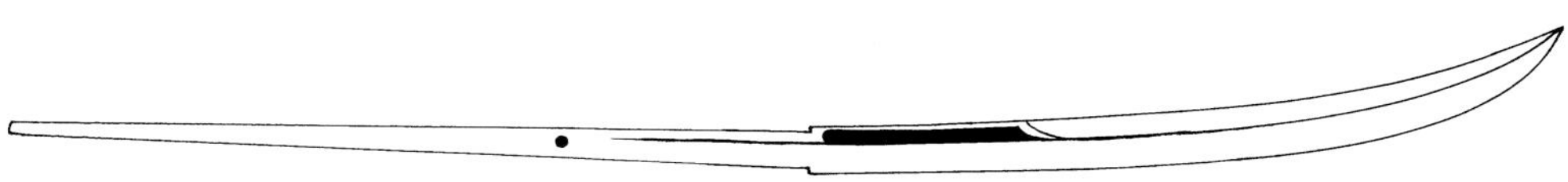

YARI.

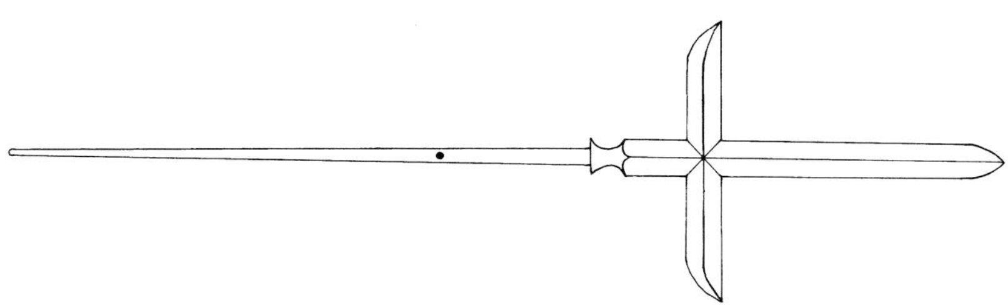

YARI.

YARI.

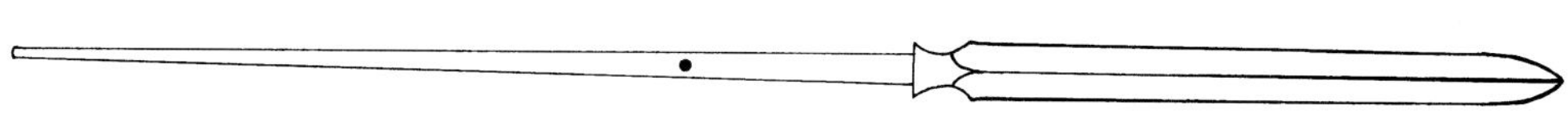

HOKO.

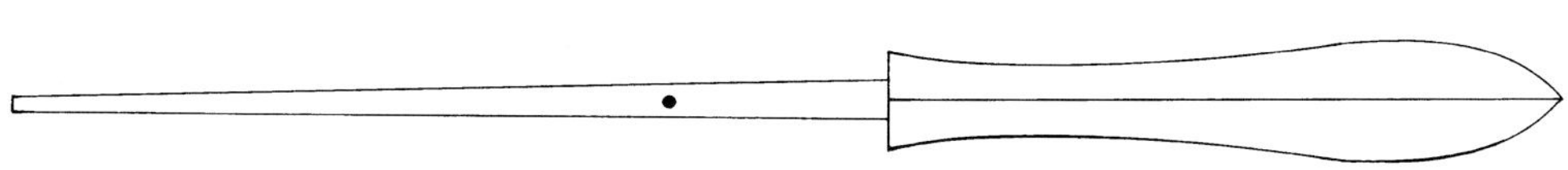

PARTS OF THE JAPANESE SWORD

Kissaki: The point.

Fukura: The sharpened cutting edge in the point area.

Boshi: The *hamon* in the point area of the sword.

Yokote: A line which is perpendicular to the cutting edge. The *yokote* defines the point area by making a clear demarcation between the point and the body of the sword.

Ko-shinogi: The part of the *shinogi* which extends through the point area of the sword.

Mitsukado: (Literally, "three corners") the point where the *yokote* line meets the *shinogi*.

Iori: A two surface back (or *mune*) of the sword. The two surfaces slope back from the sides of the sword and meet in a single sharp ridge on the top (*mune*) surface of the blade.

Mune: The top surface of the sword.

Shinogiji: The flat surface of the sword between the *shinogi* and the *mune*.

Shinogi: The ridge line running along the length of the sword parallel to the cutting edge. The thickest part of the sword is the section through the *shinogi*.

Hiraji: Basically, the *ji*, or surface of the sword above the *hamon*. On a *hira-zukuri* style sword, the *hiraji* is the area from the top of the *hamon* up to the top of the sword (the *mune*). On a *shinogi-zukuri* style blade, this is the area from the top of the *hamon* to the *shinogi*. This term is usually used to describe *hira-zukuri* type blades.

Habuchi/Nioi-guchi: The boundary between the hardened edge region (the *hamon*) and the body of the sword (the *ji*).

Hamon: The visible pattern formed by the hardened cutting edge against the body of the sword.

Ha: The cutting edge of the sword (this is the same area described by the term *hasaki*).

Hasaki: The cutting edge of the blade.

Monouchi: The portion of the blade used for cutting. This is usually the cutting edge near the point of the sword.

Machi: A notch or extension at the top of the tang of the blade. This notch is formed where the blade widens out from the tang to form the cutting edge (the *ha-machi*) or the back (*mune*) surface of the sword (*mune-machi*).

Nakago: The tang of the sword.

Yasurime: File marks.

Nakagojiri: The tip or bottom of the *nakago*.

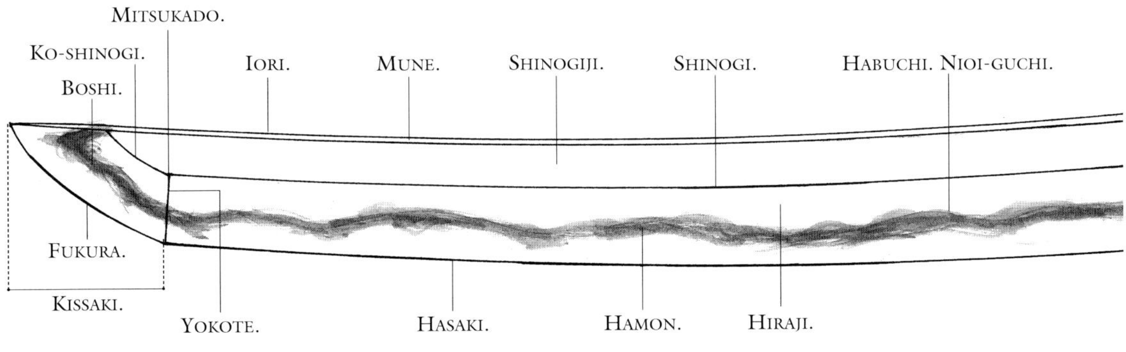

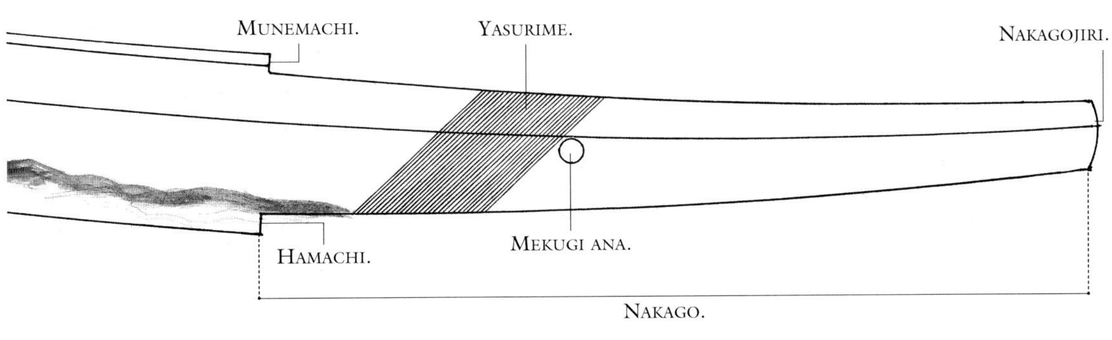

MEASURING THE JAPANESE SWORD

Nagasa/Hawatari: The length, or *nagasa*, of the sword is defined as the length of a line running from the beginning of the *munemachi* (the top of the tang) to the tip of the point.

Sori: Curvature. To determine the *sori* of a sword, a line is made to measure the length, or *nagasa*, of the sword. A perpendicular is then drawn from the line determining the *nagasa* to the *mune* of the sword. If the sword has some curvature, the length of the longest possible line between the *nagasa* line and the *mune* is called the *sori*. This will be at the point where the curvature of the sword is at its maximum.

Kasane: The thickness of the blade measured across the *mune*, or back surface. Note that the thickness of a sword can be larger at the *shinogi* than at the *mune* where the thickness is measured. There are two places where *kasane* is measured—one at the *yokote*, which is called *saki-kasane*, and the other at the *munemachi*, which is called *moto-kasane*.

Mihaba: The width of the blade measured from the cutting edge to the ridge. Since most blades taper from the base of the cutting edge (just above the tang) to the *kissaki*, there are two places where *mihaba* is measured—one at the *yokote* and the other at the *machi* (*hamachi* and *munemachi*). *Mihaba* at the *yokote* is called *saki-haba* and *mihaba* at the *machi* is called *moto-haba*.

Funbari: Tapering of the sword from the base to the point.

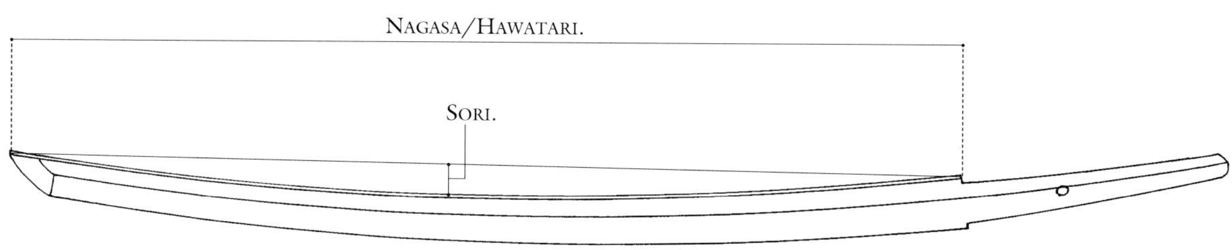

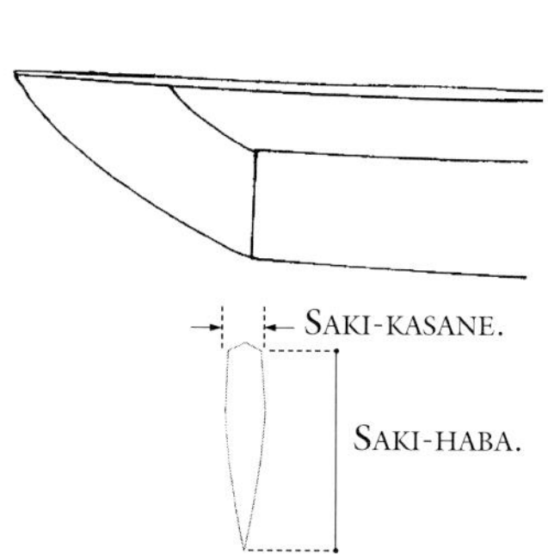

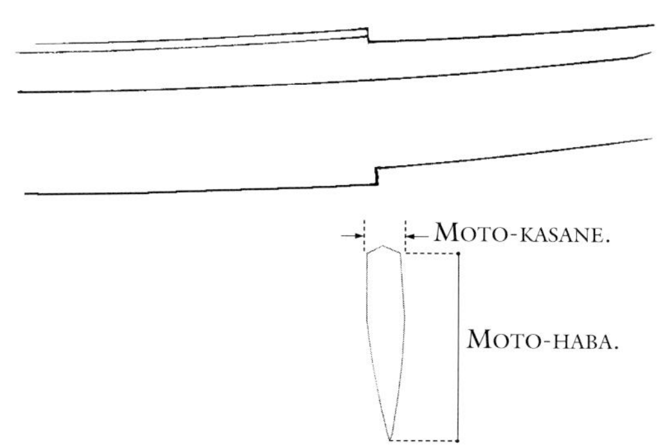

SWORD STRUCTURES (*TSUKURIKOMI*)

Hira-zukuri: The sides of the blade are flat from the edge to the back surface.

Kiriha-zukuri: The *shinogi* is very low and closer to the *ha*, or cutting edge, than it is to the top, or *mune*, surface of the sword.

Moroha-zukuri: The blade has two cutting edges with a *hamon* on both edges. *Moroha-zukuri* was developed for *tanto*.

Shinogi-zukuri: The blade has a ridge line, or *shinogi*, running along its length with the thickest part of the blade at the *shinogi*.

HIRA-ZUKURI.

KIRIHA-ZUKURI.

MOROHA-ZUKURI.

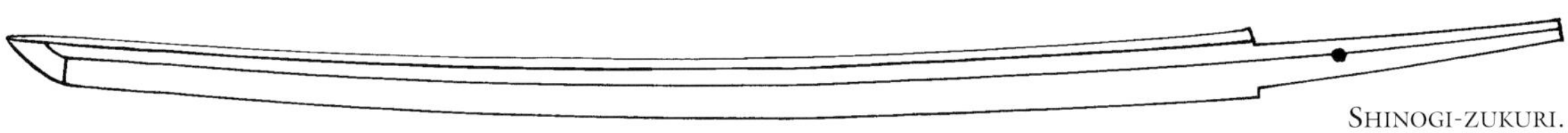

SHINOGI-ZUKURI.

TYPES OF SORI (CURVATURE)

Torii-zori: The strongest (or deepest) part of the curvature is located at the center of the sword.

Koshi-zori: The strongest (or deepest) part of the curvature is located towards the hilt of the sword rather than at the center.

Saki-zori: The strongest (or deepest) part of the curvature is located towards the point of the sword rather than at the center.

Mu-zori: There is no *sori,* or curvature, at all. The blade is straight.

Uchi-zori: There is a "reverse" curvature. The back of the blade curves towards the edge. *Uchi-zori* is the result of altering a blade or repairing a damaged blade, or overpolishing. No blades were originally made by a smith to be *uchi-zori.*

TORII-ZORI.

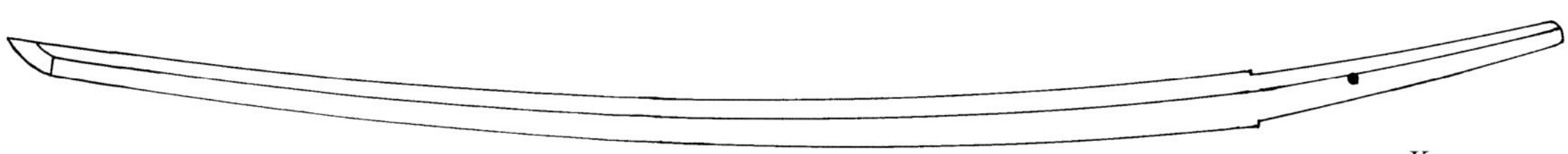

KOSHI-ZORI.

SAKI-ZORI.

MU-ZORI.

UCHI-ZORI.

TYPES OF *KISSAKI*

Ko-kissaki: A small point.

Chu-kissaki: A middle-sized or average-sized point.

O-kissaki: A very large point.

Ikubi-kissaki: A stubby, wide point (a "boar's neck" point). The point is approximately as wide as the blade and the length of the point is approximately the same as its width.

Fukura (rounded or straight): The cutting edge in the point area (the cutting edge from the *yokote* to the tip of the point). It is usually curved, but some *fukura* in the Kamakura period were straight.

KISSAKI

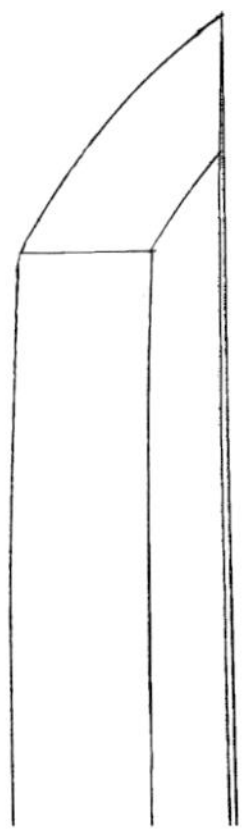

Ko-kissaki.

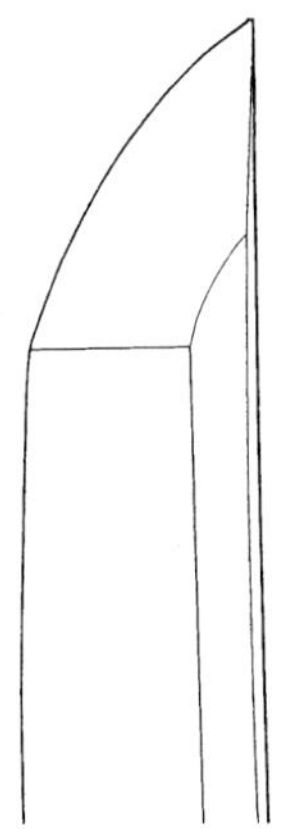

Chu-kissaki.

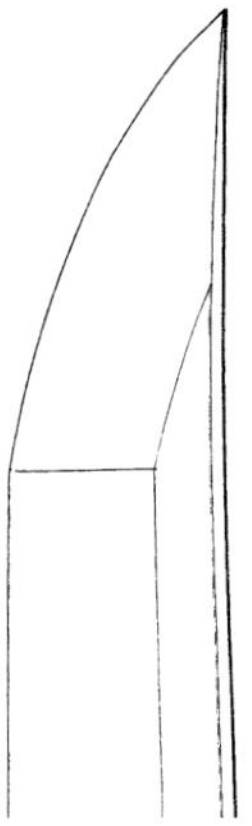

O-kissaki.

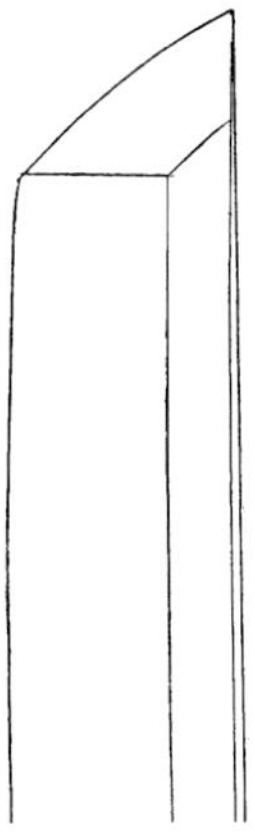

Ikubi-kissaki.

FUKURA

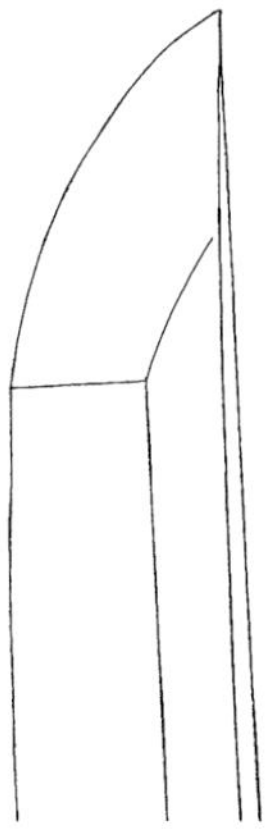

Rounded fukura.

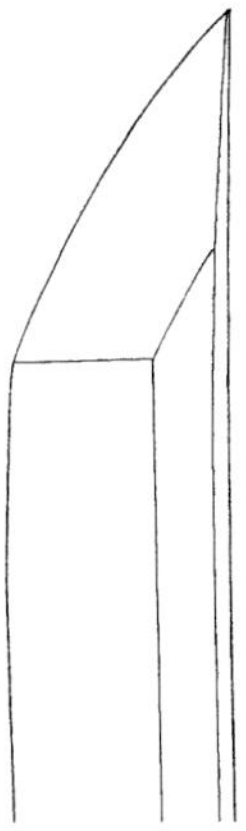

Straight fukura.

TYPES OF *MUNE*

Iori-mune: Two-sided *mune* or back surface. The two sides meet in a sharp ridge.

Mitsu-mune: Three-sided back surface. The back (upper) surface of the sword will be formed like half of a hexagon.

Maru-mune: A round *mune* or back surface, of a sword.

Shallow *oroshi*: The back surface, or *mune*, rises only a small distance above the top of the *shinogi-ji* surface.

Steep *oroshi*: The two surfaces forming the back extend well above the top of the *shinogi-ji*.

TYPES OF MUNE (RIDGE)

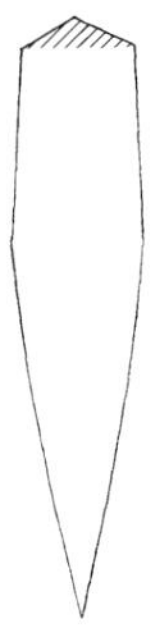

IORI-MUNE.

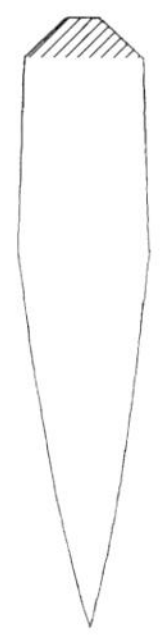

MITSU-MUNE.

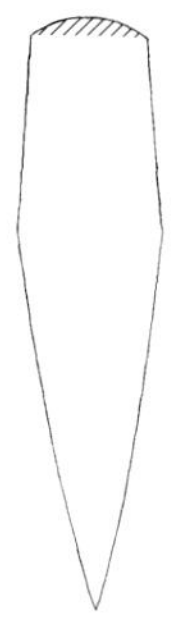

MARU-MUNE.

OROSHI

SHALLOW OROSHI.

STEEP OROSHI.

TYPES OF *SHINOGI*

Shinogi refers to the thickness of the blade from the *shinogi* line on one side of the blade to the other. If this distance is relatively large, the blade is said to be "*shinogi-ga-takai*," or a high *shinogi*. If the distance is relatively small, the blade is said to be "*shinogi-ga-hikui*," or a low *shinogi*.

Hira-niku refers to the amount of metal on the sides of the blade. "*Hira-niku-tsuku*" means that the blade sides are rounded or convex. "*Hira-niku-sukunai*" means the blade sides are flat with no convex shape or curvature. *Hira-niku* naturally decreases after repeated polishing.

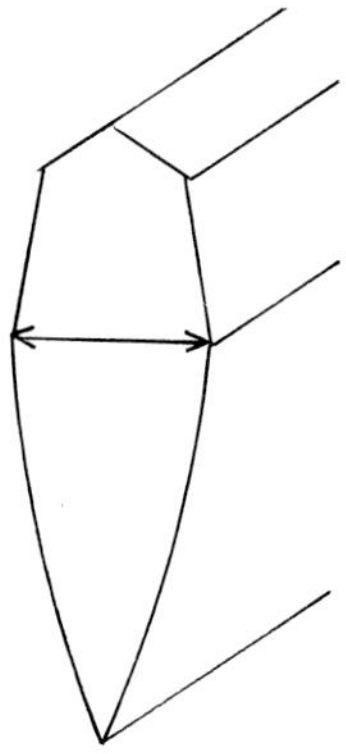

HIGH SHINOGI.

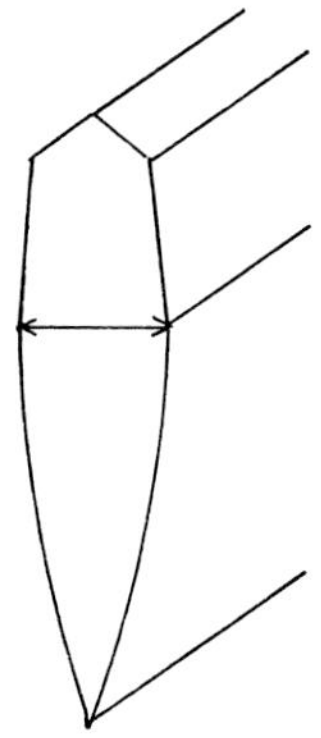

LOW SHINOGI.

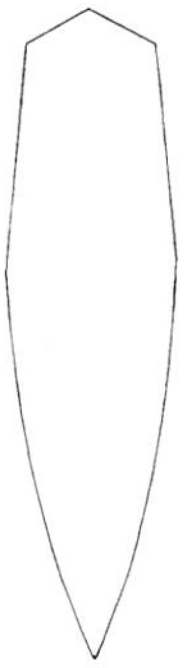

HIRA-NIKU-TSUKU.

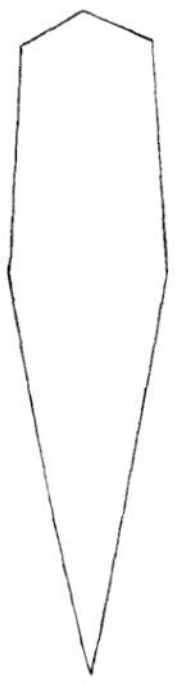

HIRA-NIKU-SUKUNAI.

TYPES OF *NAKAGOJIRI* (TANG ENDS) AND *YASURIME* (FILE-MARK PATTERNS)

Nakagojiri: The tip or bottom of the *nakago*, or tang. The *nakagojiri* refers to the shape of the *nakago* bottom on the bottom of the sword.

Kurijiri: Rounded *nakago* bottom.

Ha agari kurijiri: A rounded *nakago* bottom which is partly cut, or biased, and filed flat on the side of the *nakago* with the cutting edge.

Kiri/Ichimonji: A straight *nakago* tip that appears to be cut off.

Sujikai: Slanting downwards across the tang.

Kesho yasuri: Most of the visible *yasurime* slant downwards across the tang, but there are also horizontal and contrasting file marks filling in the space below the *machi* down to the slanting file-mark pattern that covers most of the tang.

Katte sagari: Slanting downwards, but with a very shallow angle (less than 45 degrees).

Katte agari: Slanting upwards with a shallow angle.

Higaki: Filed in a cross-hatched pattern over the tang. There are slanted file marks going in both directions, to the right and the left, as they go down the tang.

Takanoha: The file marks on the tang are filed in two different directions from the *shinogi*. On the edge side of the tang, the file marks go down from the *shinogi* to the left, while on the ridge side of the tang, the file marks go down from the *shinogi* towards the right. The effect resembles a feather with its center on the *shinogi*. *Takanoha* means "hawk feather."

TYPES OF TANG ENDS

WIDTH OF THE SHINOGI

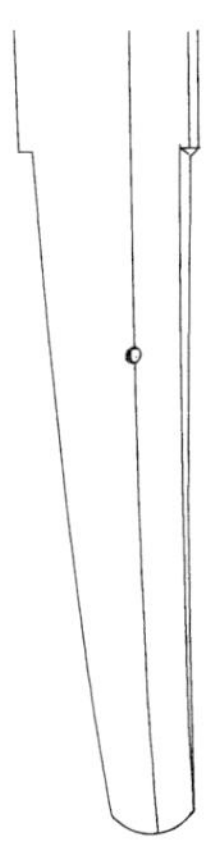
KURIJIRI.

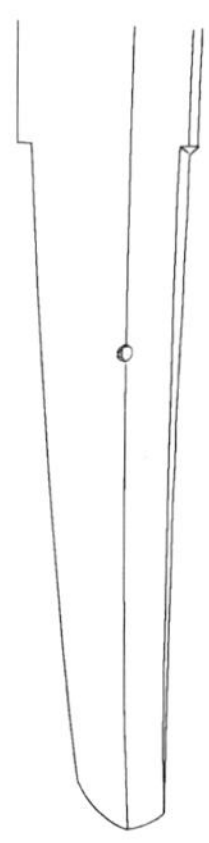
HA AGARI KURIJIRI.

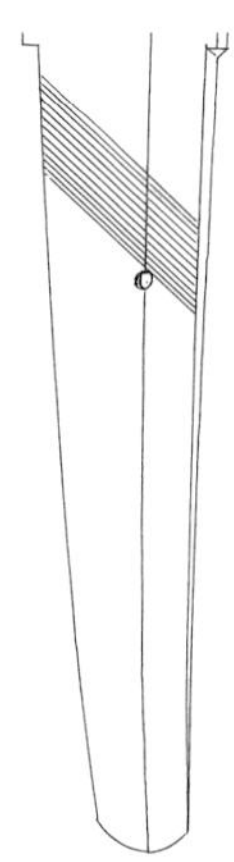
SUJIKAI.

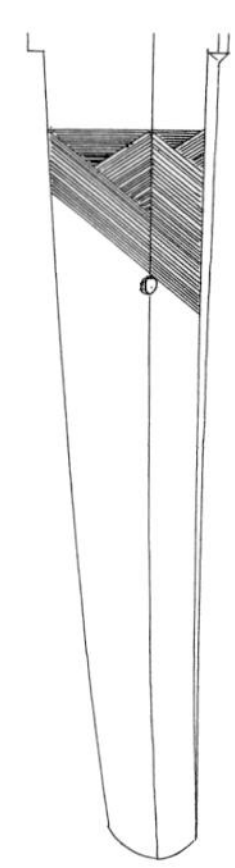
KESHO YASURI.

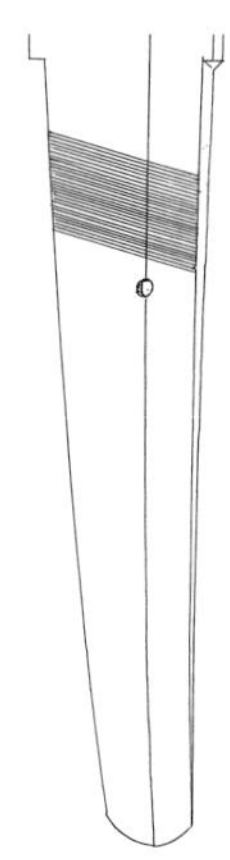
KATTE SAGARI.

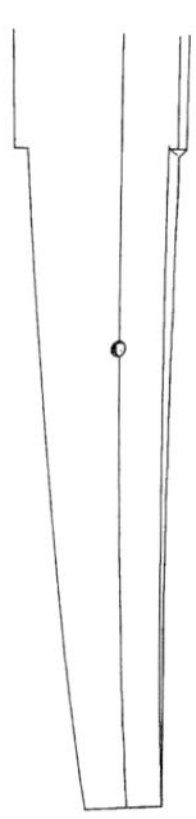
KIRI/ICHIMONJI.

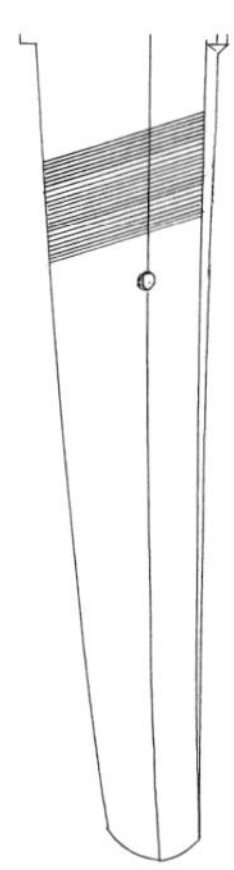
KATTE AGARI.

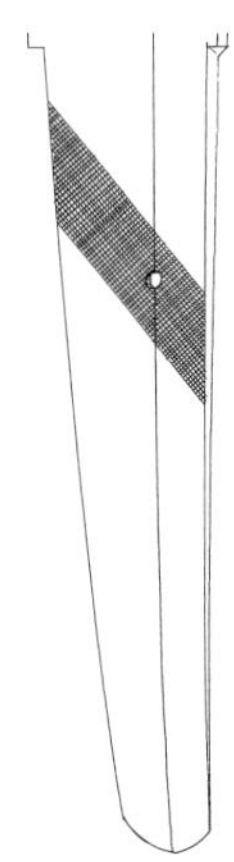
HIGAKI.

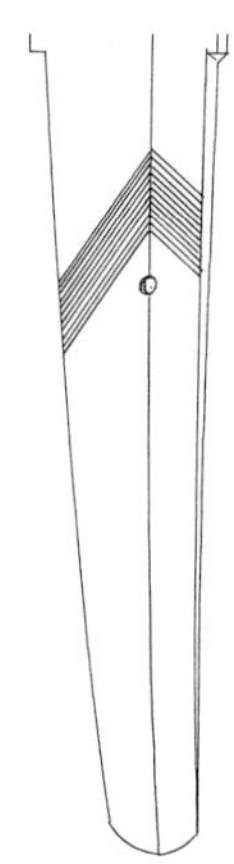
TAKANOHA.

TYPES OF GROOVES (*HI*) AND CARVINGS (*HORIMONO*)

Bo-hi : A straight groove.

Bo-bi + soe-bi: A large, straight groove (*bo-bi*) accompanied by a companion or smaller groove (*soe-bi*).

Bo-bi + tsure-bi: A large groove accompanied by a smaller companion groove. In this case, the companion groove (*soe-bi*) is longer than the large groove (*bo-bi*), and extends around the front of the large groove and towards the *mune*.

Futasuji-hi: Twin straight grooves (but not necessarily of the same length).

Kaki-toshi: The groove is cut all the way along the length of the tang.

Kaki-nagashi: The groove extends into the tang of the blade and tapers till it stops somewhere in the tang.

Kaku-dome: The groove finishes around an inch from the *machi* and stops at a squared-off flat surface, perpendicular to the axis of the blade.

Maru-dome: The end of the groove is finished with a round, even surface that forms a section of a sphere.

Goma bashi: Two parallel grooves running part way up the blade, which resemble chopsticks. The grooves may be connected at the top or bottom by additional carvings.

Suken: A straight *ken* or highly stylized sword. *Ken* is often combined with *bonji*, *sankosho* (a triple-spearhead Buddhist mallet), or a dragon.

Tokkosho: A single-spearhead Buddhist mallet.

Bonji: Sanskrit characters.

Kurikara: A dragon winding itself around a sword. The sword is usually a stylized *ken*, and the dragon is ascending towards the point of the *ken*. There are several types of *kurikara*. *Shin no kurikara*: a realistic, detailed dragon twisting around a sword. *Gyo no kurikara*: a simpler, less detailed dragon. *So no kurikara*: a very simple, stylized dragon.

Fudo-Myoo: The Buddhist deity. Fudo-Myoo is often expressed as Kaen Fudo (Fudo-Myoo framed in fire).

TYPES OF GROOVES

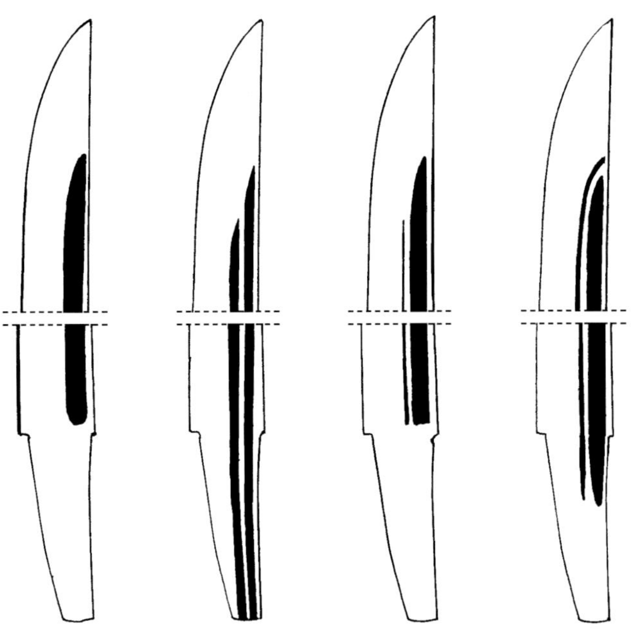

BO-HI. FUTASUJI-HI. BO-BI WITH SOE-BI. BO-BI WITH TSURE-BI.

TYPES OF GROOVE ENDS

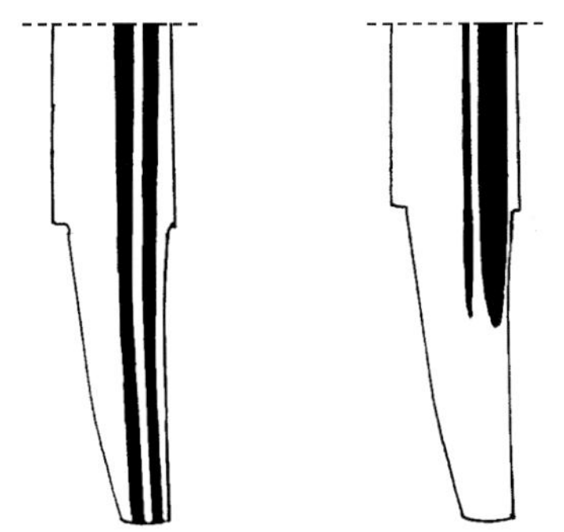

KAKI-TOSHI. KAKI-NAGASHI.

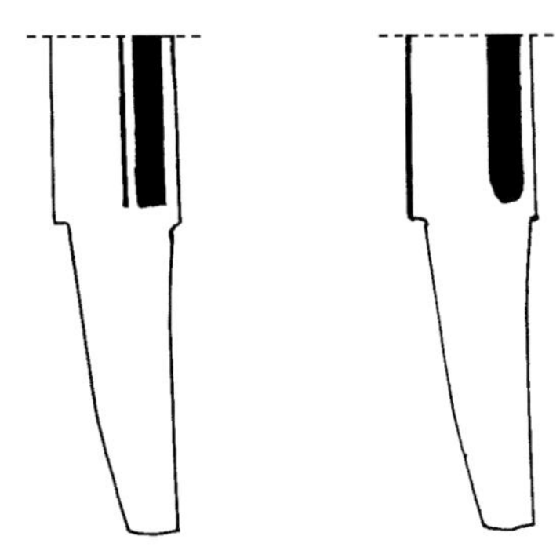

KAKU-DOME. MARU-DOME.

TYPES OF CARVINGS

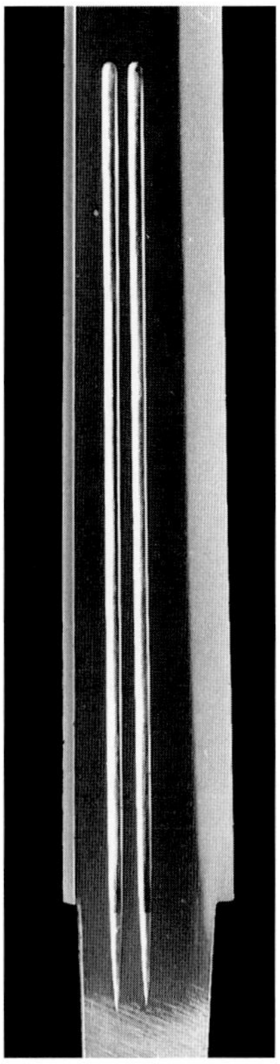

GOMA BASHI.

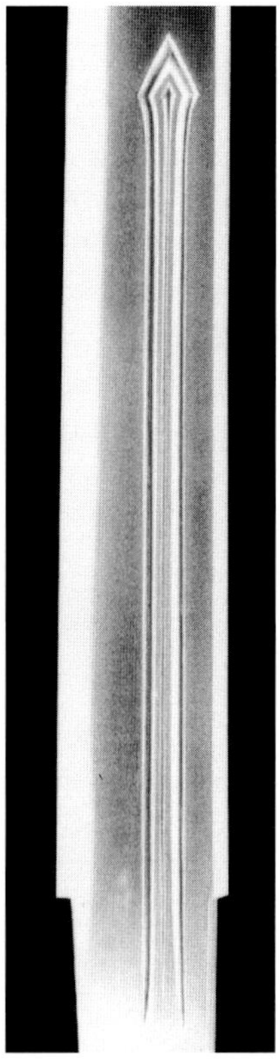

SUKEN.

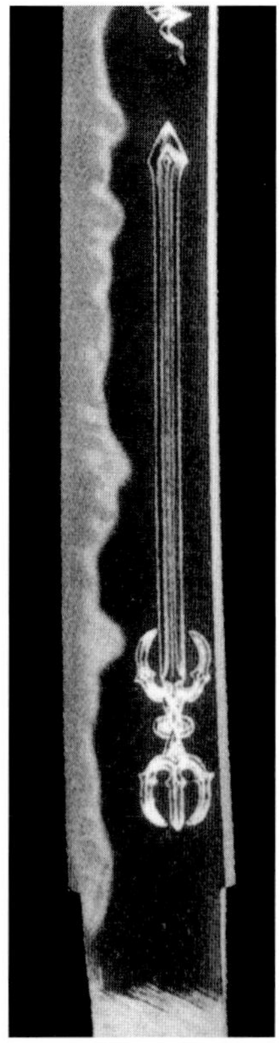

KEN WITH SANKOSHO.

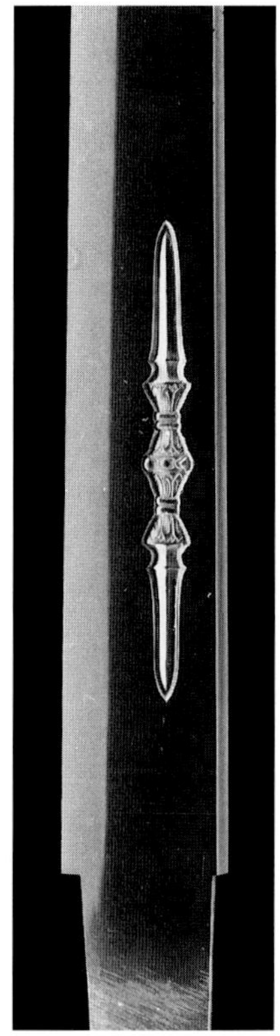

TOKKOSHO.

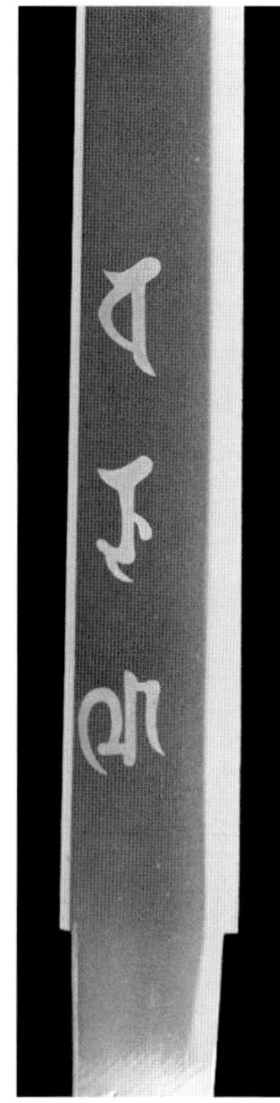

BONJI.

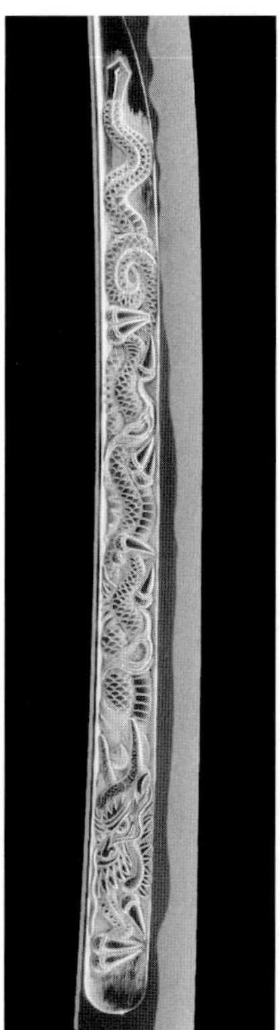

SHIN NO KURIKARA.

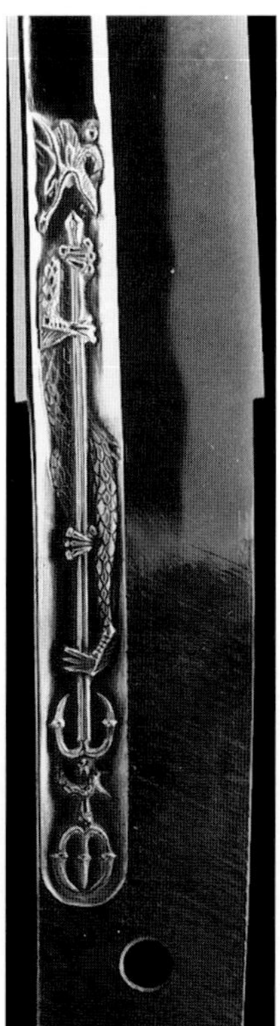

GYO NO KURIKARA.

SO NO KURIKARA.

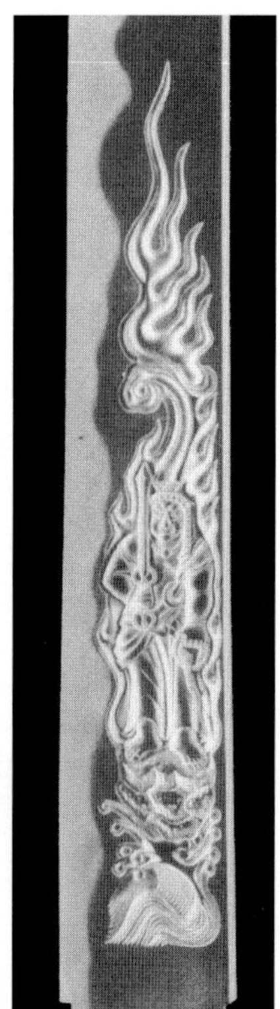

KAEN FUDO.

JIGANE and *JIHADA*

Jigane : The Japanese sword is made from a traditional steel called *tamahagane*. This steel needs to be purified, and its carbon content needs to be adjusted to meet the most essential requirements for a Japanese sword. The smith begins with *tamahagane* and adjusts its properties by working it in the forge. Thus the properties of the steel that comprises the actual blade depend on how the smith manipulated the *tamahagane* in the forge—what temperatures he used during forging, how many times he folded it, and other details. After it has been forged into a sword the steel appears very different from modern steel mass produced in a mill or a foundry. The surface is not as bright or featureless as that of modern steel, but has a definite texture, color, and distinctive appearance.

The word *jigane* has a very broad usage in sword terminology, but refers most importantly to the material of the blade; the stuff of which the sword is made. This includes the original, "raw" *tamahagane*, as well as the modified *tamahagane*, free from impurities and with a more controlled distribution of carbon. The term is also used in discussions of the sword's activities, or *hataraki*, which are effects of the unique composition and texture of the *jigane*. These are often considered to be the most aesthetically beautiful elements of the blade.

Jihada : The smith modifies the steel by repeatedly drawing it out, folding it over on itself, and drawing it out again. This process is repeated twelve to fifteen times. The continual hammering and folding results in a visible pattern on the surface of the steel, and this visible pattern is called the *jihada*. The term's usage is rather limited compared to that of *jigane*, and refers to the external, visible properties of the sword (the word "*hada*" means "skin" in Japanese). The *jihada* can have several different patterns depending on how the smith worked the steel.

Masame-hada : A straight pattern.

Mokume-hada : A pattern in which small ovals or circles are seen (similar to a fine burl-grain seen in wood). *Mokume-hada* and *itame-hada* with very fine patterns are called *ko-mokume-hada* and *ko-itame-hada*.

Itame-hada : A pattern resembling wood grain, in which scattered irregular ovals appear

Ayasugi-hada : A regular, wavy pattern.

JIGANE AND JIHADA

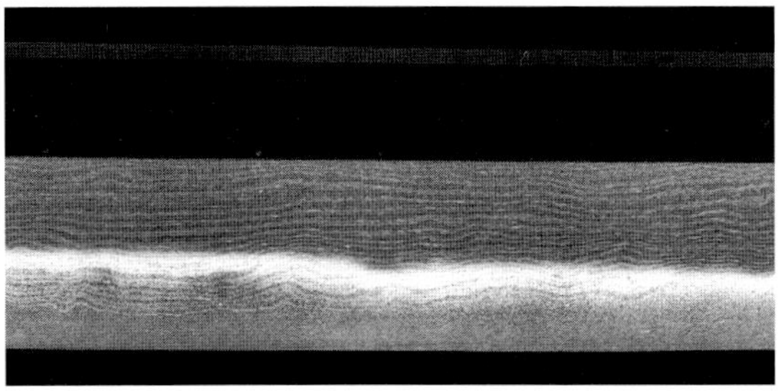

MASAME-HADA.

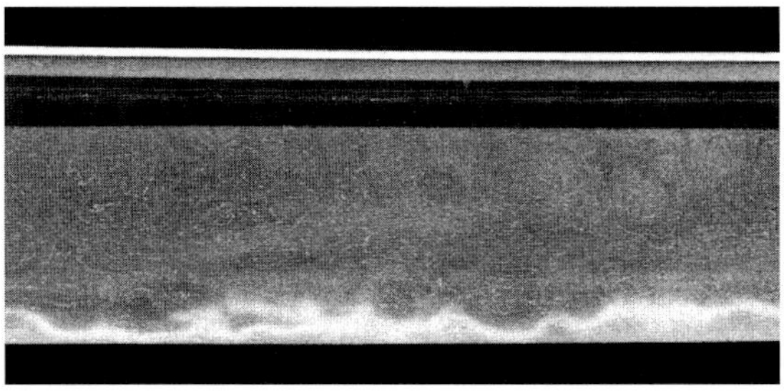

MOKUME-HADA.

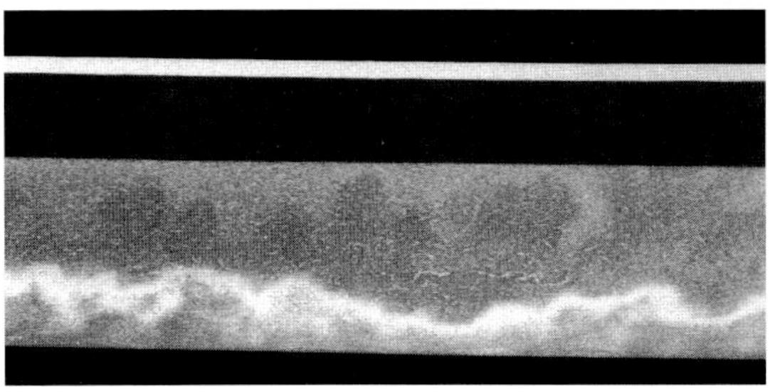

ITAME-HADA.

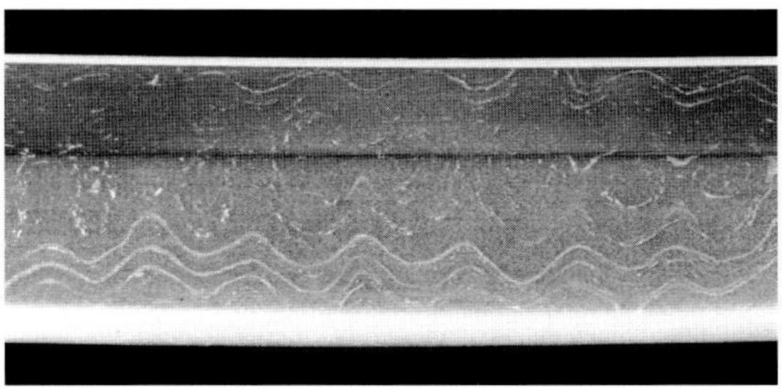

AYASUGI-HADA.

NIE AND *NIOI*

The Japanese sword is heat treated in a unique manner to harden the edge. *Yakiire* is the process of heating and then rapidly cooling the blade by plunging it into cold water. This process, when done correctly, alters the crystal structure of the steel in the edge region, producing a clear, visible temper line along the hardened edge. This line is called the "*hamon*." The *hamon* steel is composed of a crystal structure called martensite, and the line is formed by the harder martensitic crystals blending with the softer pearlite and ferrite in the body of the sword. The white line produced by the martensite is also called the "*nioi*" line, and is a feature of all swords. The *nioi* line appears as a diffuse wash of stars, and no discrete crystal particles or dots are visible. "*Nie*" are single, large martensite crystals which can be seen as discrete dots or islands. *Nie* can often be found along the *nioi* line, or above the *nioi* line in the *ji* of the sword. *Ara-nie* are very large, discrete *nie* particles or dots. *Ko-nie* are very small *nie* particles or dots that can barely be seen, but are nevertheless discernible, and so are distinct from *nioi*. *Nioi deki* means that the *hamon* boundary is composed of *nioi* with little or no *nie*. *Nie deki* means that the *hamon* contains frequent and clearly visible *nie* particles.

NIE AND NIOI

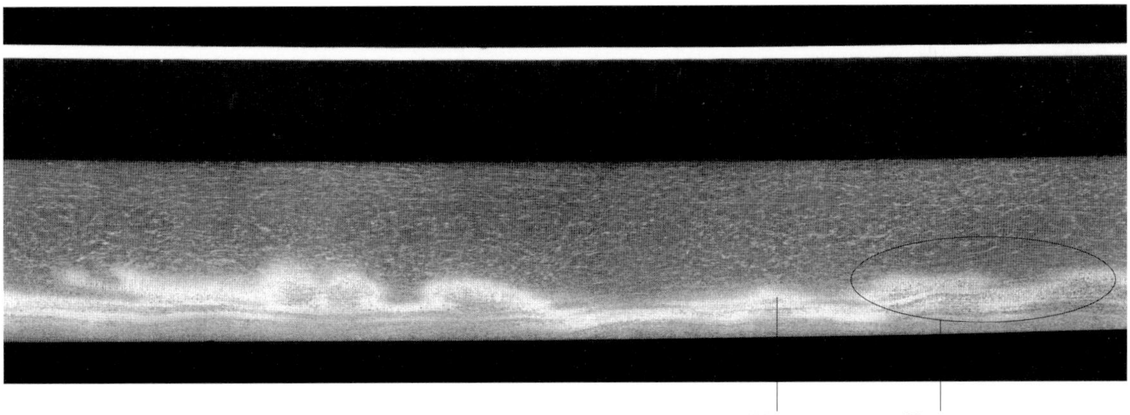

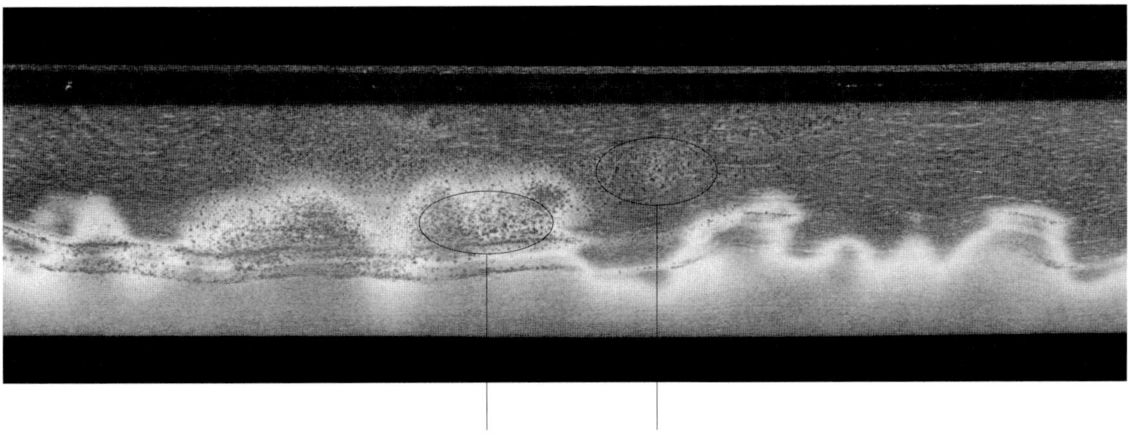

TYPES OF *HAMON*

The *hamon* is the visible, hardened edge of the Japanese sword. Its appearance will depend on what the swordsmith has done, and there are many variations and names for the different *hamon* patterns.

Suguha: A straight *hamon*. *Hoso-suguha* is a narrow, straight *hamon*; *chu-suguha* is a medium-width, straight *hamon*; and *hiro-suguha* is a wide, straight *hamon*.

Choji: A wave or loop formed by the *nioi* line defining the *hamon* which is wider at the top than at the base on the *nioi* line. A single *choji* resembles the outline of the bud of a clove blossom formed by the *nioi* line. The term "*choji*" is broad, and can be used to refer to loops or waves of varying shapes.

Juka choji: Double or multiple *choji* clustered together.

Kawazuko choji: Much wider at the top than at the base; the top of the loop can be squared off or flattened which makes the *choji* wave resemble the shape of a tadpole (*kawazuko*) when viewed from above.

O-busa choji: Larger *kawazuko choji*, distributed irregularly and clustered together along the *hamon*.

Gunome: Regular waves or loops. *Gunome* are the same width at the base and top of the loop or wave.

Kataochi gunome: The top of the waves in the *gunome* are not rounded but flat, and the whole is slightly slanted.

Gunome midare: The top of the *gunome hamon* travels irregularly over the surface of the sword (in contrast to a straight, or regular, top to the *hamon* pattern).

Notare: The outline of the *hamon* is irregular and undulating (not straight).

Toran-ba: Large, smooth waves along the *hamon* which form high peaks and deep valleys.

Hitatsura: In addition to the normal *hamon* being present, there are hardened regions (similar to the *hamon*) distributed over the sides of the blade on the *ji*.

Sanbon-sugi: Literally, "three cedars," since the *hamon* pattern displays many groups of three cone-shaped waves, or peaks, which resemble three cedar trees.

TYPES OF SUGUHA

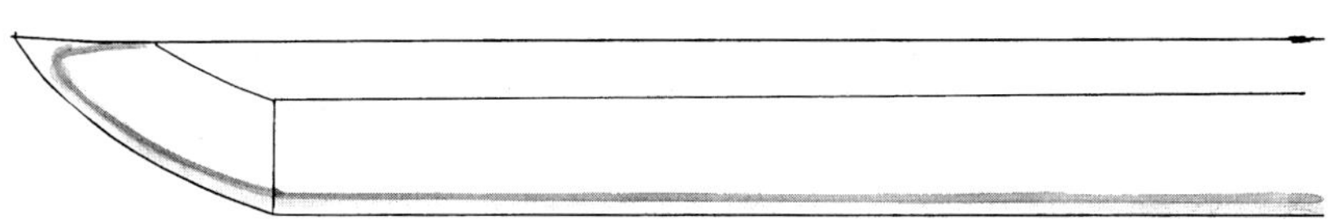

HOSO-SUGUHA.

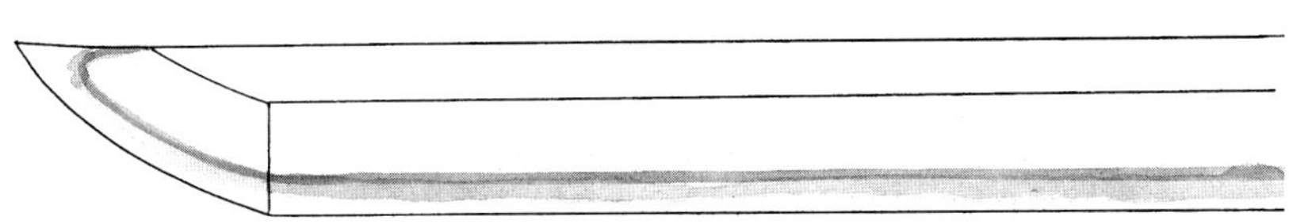

CHU-SUGUHA.

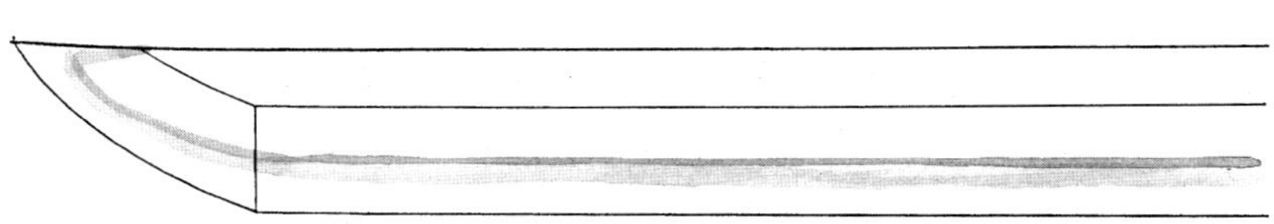

HIRO-SUGUHA.

VARIOUS HAMON

Choji with kawazuko choji.

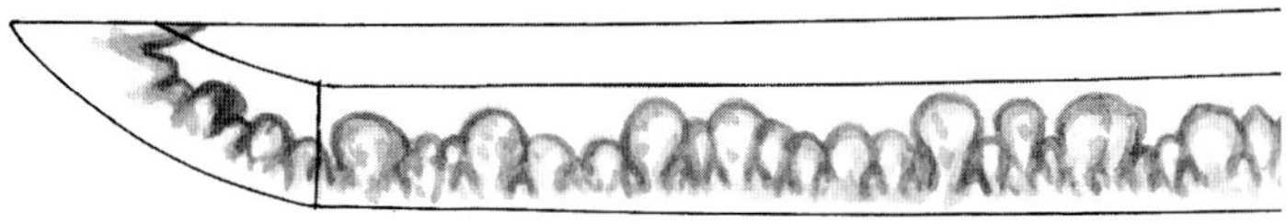

Juka-choji with o-busa choji.

Saka choji/slanted choji.

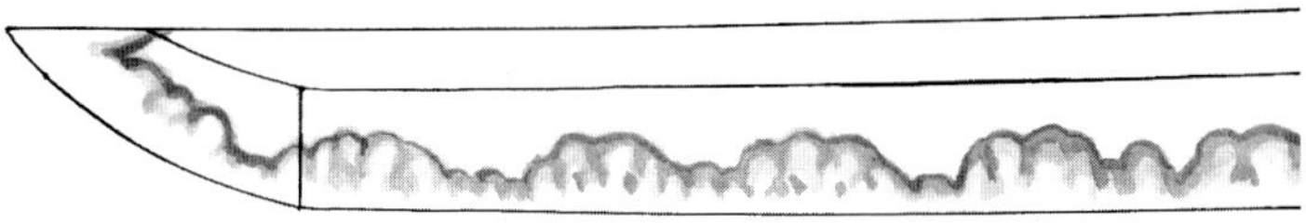

Gunome choji.

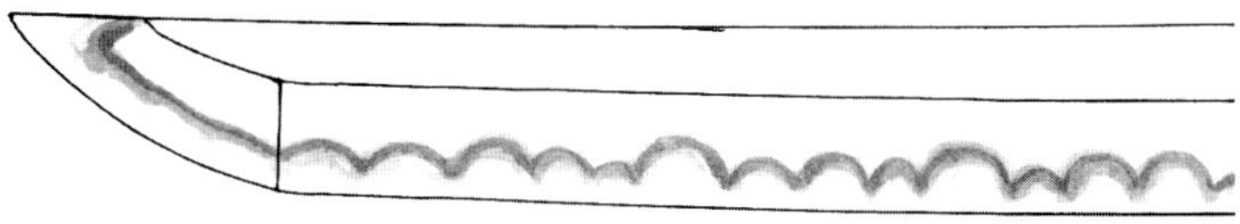

Gunome.

Kataochi gunome.

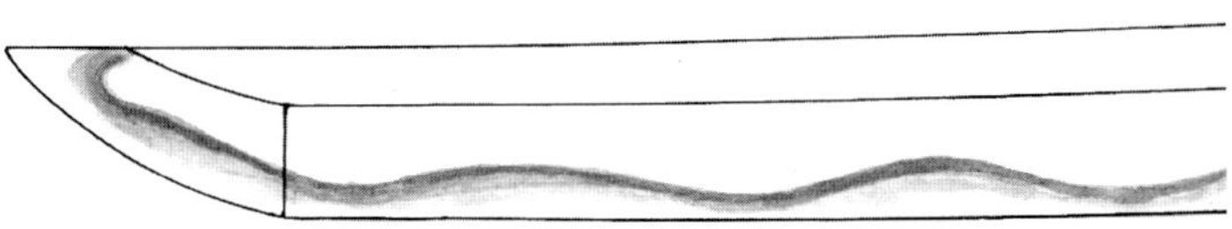

Notare.

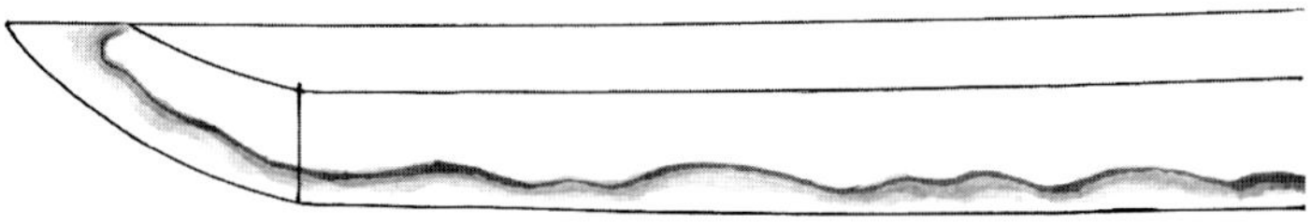

Ko-notare.

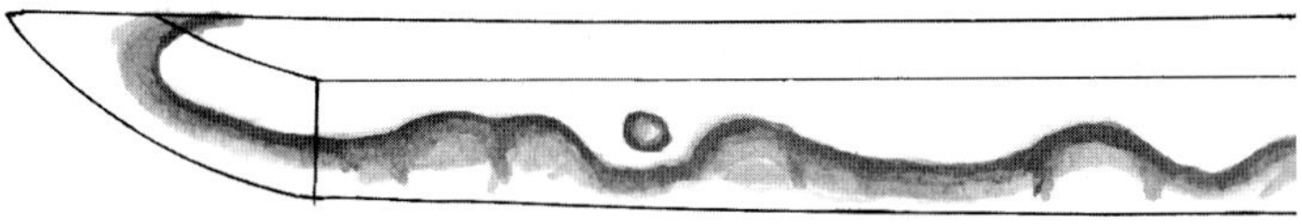

Toran-ba.

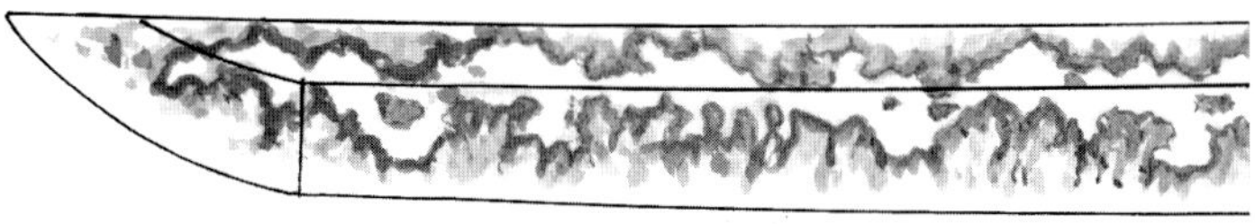

Hitatsura.

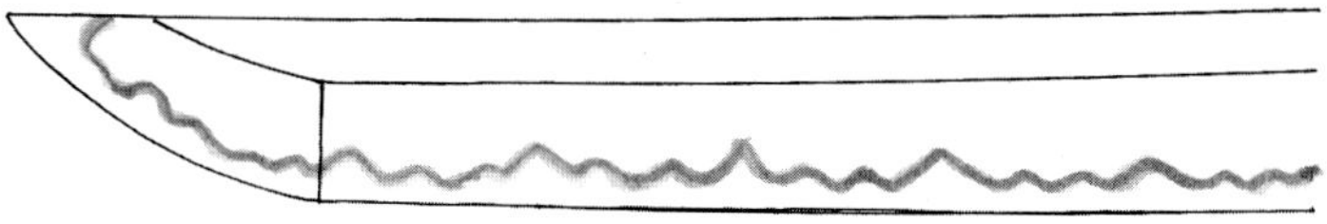

Sanbon-sugi.

HATARAKI (ACTIVITY)

There are various visible effects like lines, streaks, dots, and hazy reflection in addition to the grain patterns and *hamon* on the surface of a blade. These features seen in the *ji* and the *hamon* are called "*hataraki*," or activity, the presence of which is usually considered a crucial element in sword appreciation.

HATARAKI WITHIN THE *JI*

Utsuri: A whitish cast or coloring which appears on the *ji* of the sword, somewhere above the *hamon*. It means "reflection" and is described as a reflection of the *hamon* in the *ji*, but it is not composed of hardened martensitic steel like the *hamon*. *Utsuri* can have many shapes and forms like that of the *hamon*.

Bo utsuri: A straight line of *utsuri* on the *ji*.

Midare utsuri: A wavy, or undulating, *utsuri* on the *ji*.

Shirake utsuri: Irregular *utsuri* over the *ji*, with no defined pattern.

Chikei: Hardened or bright streaks or lines of steel seen in the *ji*. *Chikei* has a higher carbon content than the rest of the blade.

Yubashiri: Hardened spots or regions seen in the *ji* that are composed of martensitic steel like the *hamon* (the *yubashiri*, however, are not related to the *hamon*).

HATARAKI WITHIN THE *HAMON*

Ashi: Lines projecting down from the *hamon* boundary towards the edge.

Yo: A variation of ashi, which looks like a spot inside the hamon.

Nijuba: A double line defining the *hamon*. Instead of a single *nioi* line marking the *hamon* boundary, there are two closely spaced parallel lines. The *nijuba* might only cover part of the length of the *hamon*.

Hotsure: The *nioi* line defining the *hamon* boundary which looks like a frayed or unraveling thread rather than a single, dense line.

Kinsuji: Irregular, clear lines inside the *hamon* which have a higher carbon content than the rest of the blade.

Inazuma: Metallurgically identical to *kinsuji*, but its shape is different, resembling lightning bolts.

Sunagashi: Lines inside the *hamon* that resemble grains of sand. When the pattern is clear, the carbon content is lower than the rest of the blade, and when it is weak or faint, the carbon content is higher.

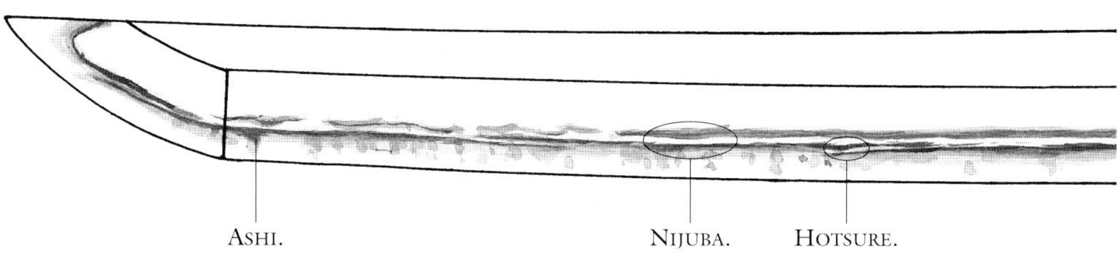

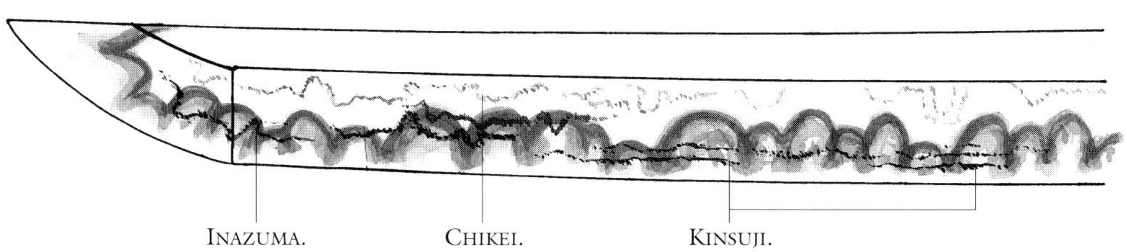

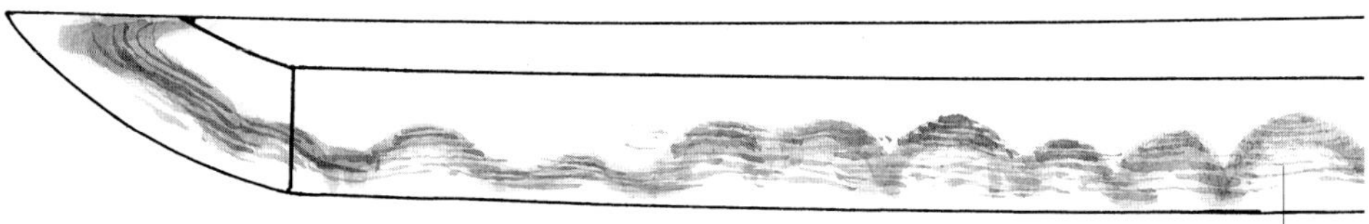

Sunagashi.

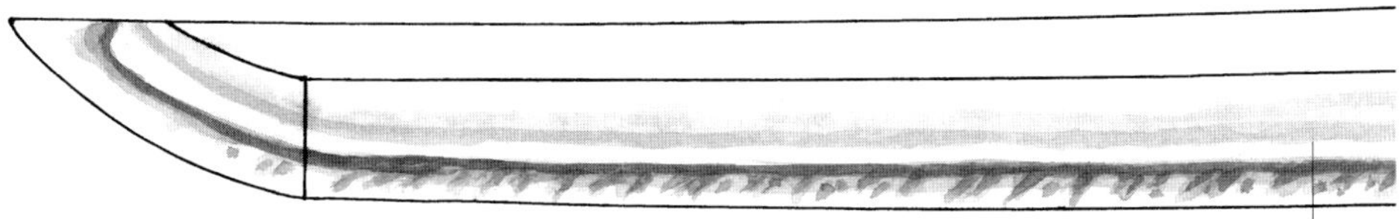

Bo utsuri.

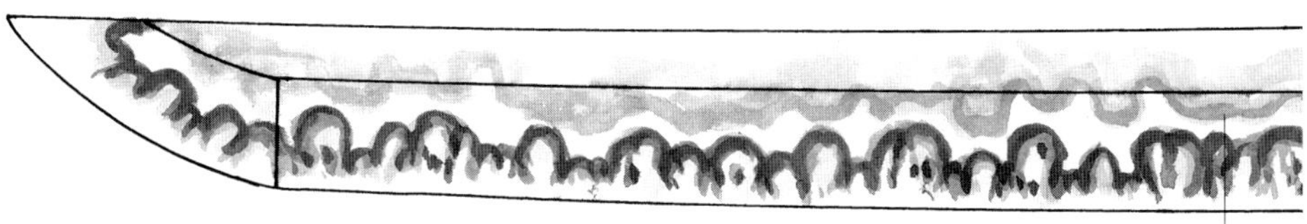

Midare utsuri.

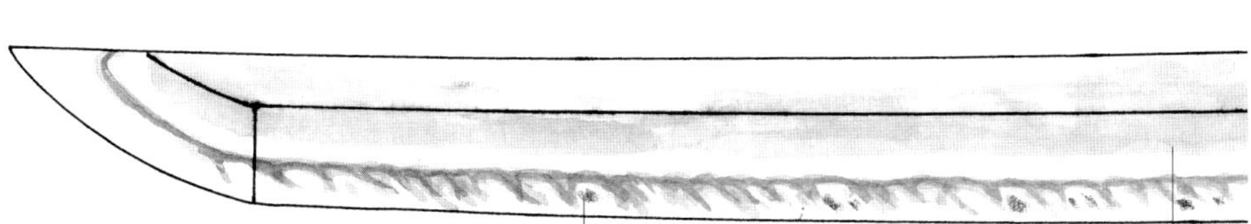

Yo.

Shirake utsuri.

Yubashiri.

TYPES of *BOSHI*

The *boshi* is the *hamon* in the point area of the sword. There are several terms used to describe the *boshi*.

Ko-maru: The *hamon* enters the point area, follows the point area in a straight line, and forms a small circle as it travels up to the unsharpened top surface, of the point.

O-maru: The *boshi* is straight as it enters the point area and then curves around and goes to the top (unsharpened) portion of the point. The turn, or curve, the *boshi* makes as it goes around the point area is a large circle.

Tsukiage: The *boshi* goes to the point area, and forms a sharp point as it turns back towards the tang.

Midare-komi: The *irregular* pattern of the *hamon* continues into the *boshi*. There are various *kaeri* (turn back) in this type.

Yakitsume: The *boshi* goes forward towards the point of the blade, and stops when it reaches the *mune*, or back surface, of the blade; there is no turning or circling back of the *nioi* line towards the tang.

Hakikake: As the hamon enters the point area, it is not defined by a single, clear *nioi* line, but instead by a series of almost parallel *nioi* lines resembling a broom stroke.

Jizo: The *boshi* forms a round, head-like shape (resembling the profile of a common stone figure called a "*Jizo*") as it turns back towards the tang.

Ichimai: The point is completely hardened because the *hamon* turns back before it reaches the point, so there is no visible *nioi* line or *hamon* inside the point area.

Sansaku boshi: This is the shape of the *boshi* made by the three swordsmiths Osafune Nagamitsu, Kagemitsu, and Sanenaga. Once the *hamon* enters the point area, it becomes a straight line and goes towards the tip of the point. Just before the *boshi* (*hamon*) reaches the point, it forms a small circle (*ko-maru*) and then turns back towards the tang. And the *boshi* only turns back a very short distance, however, remaining inside the point area.

TYPES OF BOSHI

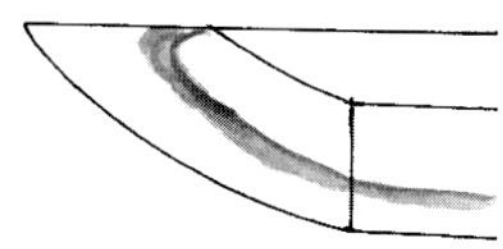

KO-MARU.

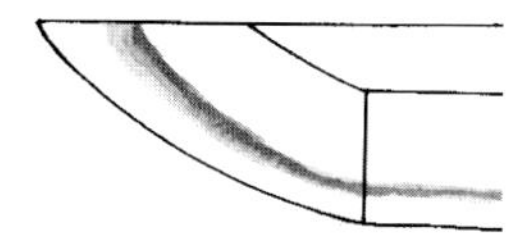

YAKITSUME.

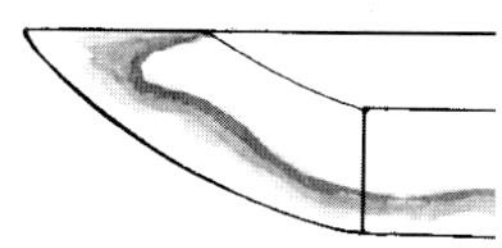

NOTARE KOMI KO-MARU.

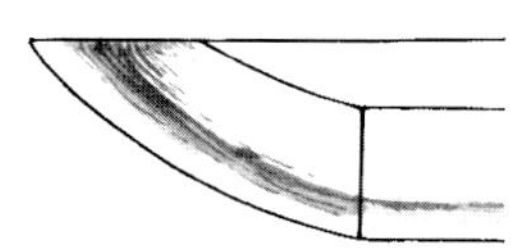

HAKIKAKE.

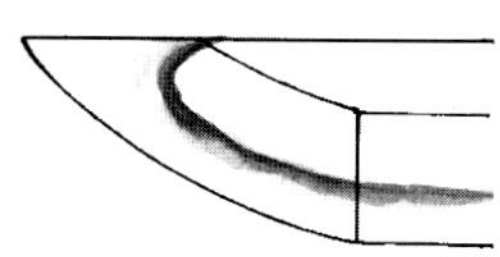

O-MARU.

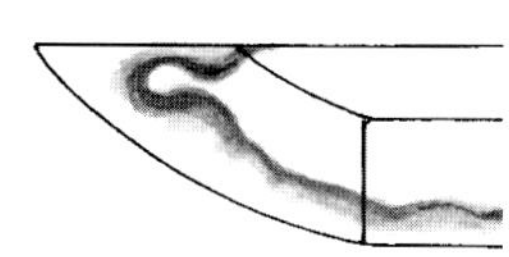

JIZO.

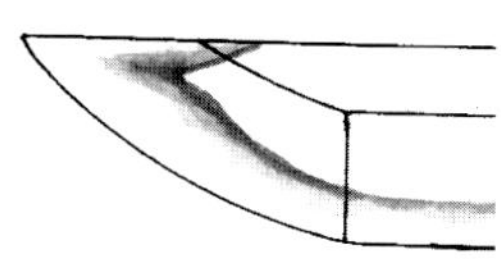

TSUKIAGE.

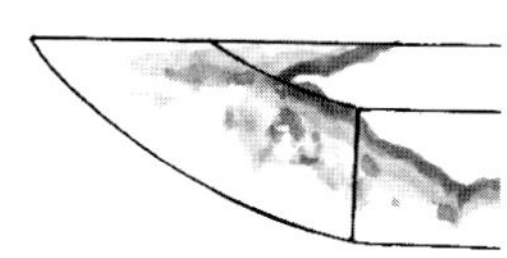

ICHIMAI.

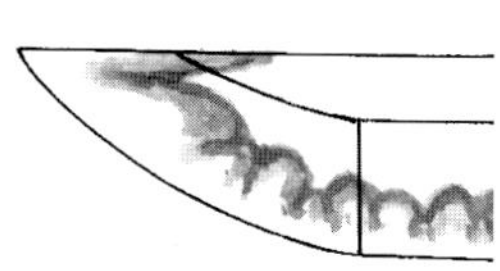

MIDARE-KOMI TSUKIAGE.

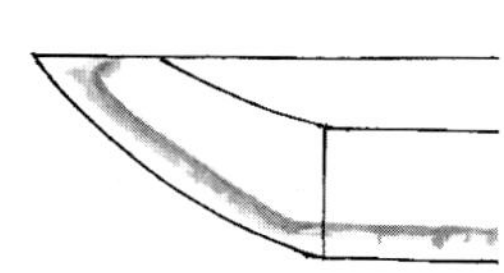

SANSAKU.

A detail of a shinto sword with a straight hamon. A wood-grain-patterned jihada (itame-hada) with a circle pattern (mokume-hada) is visible in the ji.

ACKNOWLEDGMENTS

For some of the authors of this book (the Kapps and Yoshihara), this is the third volume they have published with Kodansha International. With all three efforts, it has been a pleasure to work with Shigeyoshi Suzuki and his colleagues at Kodansha International. The authors greatly appreciate all of the thought and work that went into producing this series of sword books.

We particularly wish to acknowledge the efforts and enthusiasm of Shigeyoshi Suzuki in helping to acquire and assemble the material used in this book. He has been a wise guiding light. We also wish to express our gratitude to Barry Lancet for his advice and suggestions, and to Masumi Akiyama for a design that enlivens the subject matter and has impressed us all.

Setsuo Takaiwa
Yoshindo Yoshihara
Leon and Hiroko Kapp

BIBLIOGRAPHY

Fujishiro, Okisato. "Nihonto Tanren Kai ni tsuite" (On Nihonto Tanren Kai). *Token Bijutsu* (Sword Art), vols. 288–9, Tokyo: Nihon Bijutsu Token Hozon Kyokai, 1981.

Fuller, Richard; and Gregory, Ron. *Military Swords of Japan, 1868–1945*. London: Arms and Armor Press, 1986.

Fuller, Richard; and Gregory, Ron. *Japanese Military and Civil Swords and Dirks*. Charlottesville, VA.: Howell Press, Inc., 1997.

Kapp, Leon: Kapp, Hiroko; and Yoshihara, Yoshindo. *The Craft of the Japanese Sword*. Tokyo: Kodansha International, 1987.

Kapp, Leon and Hiroko; and Yoshindo Yoshihara. *Modern Japanese Swords and Swordsmiths: From 1868 to Present*. Tokyo: Kodansha International, 2002.

Kishida, Tom. *Yasukuni-to: Dento to Bi no Kyokuchi* (Yasukuni Swords: Tradition and Ideal Beauty). Tokyo: Yuzankaku Publishing Co., 1998.

Kishida, Tom. *Yasukuni Tosho* (Yasukuni Swordsmiths). Tokyo: Tom Kishida, Tokyo. 1994.

Nagayama, Kokan. *Token Kantei Tokuhon* (*The Connoisseur's Book of Japanese Swords*; (Eng. trans. 1997, Kodansha International). Tokyo: Nagayama Bijutsu Token Kenma Jo, 1996.

Sato, Kanzan. *Token* (*The Japanese Sword*; Eng. trans. 1983, Kodansha International). Tokyo: Shibundo, 1966.

Sato, Kanzan. "Sengo Tanto Kai no Ayumi" (Postwar Sword Forging Paths). *Gendaito Meisaku Zukan* (A Photographic Introduction to the Masterpieces of Modern Japanese Swords). Tokyo: Token Shunju Shinbun Sha, 1977.

Tsuchiko, Tamio. *Kurihara Hikosaburo Akihide Zen Kiroku: Nihonto o Nido Yomigaeraseta Otoko* (A Complete Record of the Life of Kurihara Hikosaburo Akihide: The Man who Saved Japanese Swords Twice). Tokyo: Kurihara Hikosaburo Denki Kanko Kai, 2000.

Uchiyama, Takashi. "Nihon Gendaito Shoshi" (A Short History of Modern Swords). *Token Bijutsu* (Sword Art), vols. 146–9, 151–3, 166–7, Tokyo: Nihon Bijutsu Token Hozon Kyokai, 1969–70.

Yumoto, John. *The Samurai Sword*. Tokyo: Charles E. Tuttle, 1958.

INDEX

(英文版) 日本刀と研ぎ
The Art of Japanese Sword Polishing

2006年2月24日　第1刷発行

著　者　高岩節夫、吉原義人、レオン・カップ、啓子・カップ
撮　影　トム・岸田、吉原義一
発行者　富田 充
発行所　講談社インターナショナル株式会社
〒112-8652　東京都文京区音羽1-17-14
電話　03-3944-6493 (編集部)
03-3944-6492 (マーケティング部・業務部)
ホームページ　www.kodansha-intl.com

印刷・製本所　大日本印刷株式会社

Printed in Japan
ISBN4-7700-2494-0